This is the first systematic account of the pinnipeds to be published [illegible] fifty years. Drawing [illegible] knowl-

Seals Sea Lions and Walruses

A Review of the Pinnipedia

by
VICTOR B. SCHEFFER

STANFORD UNIVERSITY PRESS
Stanford, California

LONDON: OXFORD UNIVERSITY PRESS
1958

STANFORD UNIVERSITY PRESS, STANFORD, CALIFORNIA
LONDON: OXFORD UNIVERSITY PRESS

LIBRARY OF CONGRESS CATALOG CARD NUMBER: 58-7844
PRINTED IN THE UNITED STATES OF AMERICA
PUBLISHED APRIL 24, 1958

PREFATORY NOTE

The writing of this review was financed by the National Science Foundation while the author was a member of the graduate faculty, Colorado State University, Fort Collins.

VICTOR B. SCHEFFER, *Biologist*
U.S. Fish and Wildlife Service

Seattle, Washington

CONTENTS

TABLES

FIGURES

PLATES

Following page 150

For scientific names of pinnipeds mentioned in the plates, see table 1. Unless otherwise noted, illustrations are from photographs by the author, courtesy U.S. Fish and Wildlife Service.

SEALS, SEA LIONS, AND WALRUSES

INTRODUCTION

Of the group of animals now known as the Pinnipedia (seals, sea lions, and walruses), Linnaeus wrote two hundred years ago:

"This is a dirty, curious, quarrelsome tribe, easily tamed, and polygamous; flesh succulent and tender; fat and skin useful. They inhabit and swim under water and crawl on land with difficulty because of their retracted fore-feet and united hind-feet; feed on fish and marine productions, and swallow stones to prevent hunger, by distending the stomach" (Turton's translation of 1806).

The pinnipeds of the world were reviewed by Allen in 1880 and were catalogued by Trouessart in 1897–1905. In the twentieth century, man's knowledge of life in the sea—especially in the polar regions—and his understanding of the processes of animal evolution have expanded greatly. Thus it is appropriate to examine in a fresh light the pattern of distribution and variation displayed by the pinnipeds.

These animals are presently classified in 20 genera, 31 species, and 16 subspecies (a total of 47 trivial names). Collectively they make up a world population of perhaps 15 to 25 million individuals (see table 1).

TABLE 1. ESTIMATED WORLD POPULATIONS (STOCKS) OF PINNIPEDS*

Species and Subspecies	Vernacular Name	Estimated Numbers	
	Family OTARIIDAE		
Otaria byronia (South America)	South American sea lion	300,000 to	500,000
Otaria byronia (Falkland Islands)	South American sea lion	400,000 to	500,000
Eumetopias jubata	Steller sea lion	60,000 to	150,000
Zalophus californianus californianus	California sea lion	50,000 to	100,000
Zalophus californianus japonicus	California sea lion	200 to	500
Zalophus californianus wollebaeki	California sea lion	20,000 to	50,000
Neophoca cinerea	Australian sea lion	2,000 to	10,000
Neophoca hookeri	New Zealand sea lion	10,000 to	50,000
Arctocephalus pusillus	South African fur seal	300,000 to	600,000
Arctocephalus forsteri	New Zealand fur seal	5,000 to	20,000
Arctocephalus doriferus	Australian fur seal	5,000 to	20,000

TABLE 1–*Continued,* Family OTARIIDAE

Arctocephalus gazella	Kerguelen fur seal	13,000 to	15,000
Arctocephalus australis australis	South American fur seal	25,000 to	40,000
Arctocephalus australis gracilis	South American fur seal	80,000 to	200,000
Arctocephalus australis galapagoensis	South American fur seal	100 to	500
Arctocephalus philippii philippii	Philippi fur seal	? 0 to	? 0
Arctocephalus philippii townsendi	Philippi fur seal	200 to	500
Callorhinus ursinus (Pribilof Islands)	northern fur seal	1,500,000 to	1,800,000
Callorhinus ursinus (Commander Islands)	northern fur seal	40,000 to	60,000
Callorhinus ursinus (Robben Island)	northern fur seal	40,000 to	60,000
Total Otariidae		2,850,500 to	4,176,500
	Family ODOBENIDAE		
Odobenus rosmarus rosmarus (Atlantic)	walrus	20,000 to	40,000
Odobenus rosmarus divergens (Pacific)	walrus	25,000 to	50,000
Total Odobenidae		45,000 to	90,000
	Family PHOCIDAE		
Phoca vitulina vitulina	harbor seal	40,000 to	100,000
Phoca vitulina concolor	harbor seal	40,000 to	100,000
Phoca vitulina mellonae	harbor seal	500 to	1,000
Phoca vitulina richardi	harbor seal	50,000 to	200,000
Phoca vitulina largha	harbor seal	20,000 to	50,000
Pusa hispida hispida	ringed seal	2,000,000 to	5,000,000
Pusa hispida ochotensis	ringed seal	200,000 to	500,000
Pusa hispida krascheninikovi	ringed seal	50,000 to	250,000
Pusa hispida botnica	ringed seal	10,000 to	50,000
Pusa hispida ladogensis	ringed seal	5,000 to	10,000
Pusa hispida saimensis	ringed seal	2,000 to	5,000
Pusa sibirica	Baikal seal	40,000 to	100,000
Pusa caspica	Caspian seal	800,000 to	1,500,000
Histriophoca fasciata	ribbon seal	20,000 to	50,000
Pagophilus groenlandicus (White Sea)	harp seal	1,000,000 to	1,500,000

TABLE 1—*Continued,* Family PHOCIDAE

Pagophilus groenlandicus (Jan Mayen)	harp seal	500,000 to 1,000,000
Pagophilus groenlandicus (Newfoundland)	harp seal	3,000,000 to 4,500,000
Halichoerus grypus	grey seal	25,000 to 50,000
Erignathus barbatus barbatus	bearded seal	50,000 to 100,000
Erignathus barbatus nauticus	bearded seal	25,000 to 50,000
Monachus monachus	Mediterranean monk seal	1,000 to 5,000
Monachus tropicalis	Caribbean monk seal	? 0 to ? 0
Monachus schauinslandi	Hawaiian monk seal	1,000 to 1,500
Lobodon carcinophagus	crabeater seal	2,000,000 to 5,000,000
Ommatophoca rossi	Ross seal	20,000 to 50,000
Hydrurga leptonyx	leopard seal	100,000 to 300,000
Leptonychotes weddelli	Weddell seal	200,000 to 500,000
Cystophora cristata	hooded seal	300,000 to 500,000
Mirounga leonina (Falkland area)	southern elephant seal	250,000 to 400,000
Mirounga leonina (Macquarie Islands)	southern elephant seal	50,000 to 100,000
Mirounga leonina (Heard Island)	southern elephant seal	30,000 to 60,000
Mirounga leonina (elsewhere)	southern elephant seal	50,000 to 100,000
Mirounga angustirostris	northern elephant seal	8,000 to 10,000
Total Phocidae		10,887,500 to 22,142,500
Total Pinnipedia		13,783,000 to 26,409,000

* Minimum and maximum estimates are given for each species and subspecies. The width of the gap between minimum and maximum is an indication of the author's confidence in the estimate (the wider the less confidence). Source material will be found in the systematic account starting on p. 52.

Under pressure of civilization the composition of the population is changing. Thus the Philippi fur seal is disappearing (or may in fact have gone entirely) from the fauna of South America. The fur seals of the Galapagos Islands and Isla de Guadalupe number fewer than a thousand. Should any Caribbean monk seals remain alive, the location of their breeding site is a secret known to few. The Japanese race of California sea lion is now represented by perhaps 200 to 500 individuals. The numbers of walrus and hooded seal are, in the opinion of leading biologists, steadily falling. It is not the purpose of the writer, however, to call attention to conservation needs but to point to certain relationships—zoological and geographical—among the Pinnipedia. The scope of the present work is indicated by the table of contents.

The writer is deeply grateful to The National Science Foundation for research grant NSF- G 2712 and to Colorado State University for administering the grant; to the British Museum (Natural History), Scott Polar Research Institute, Cambridge University, Smithsonian Institution (National Museum), and U.S. Department of the Interior (Board on Geographic Names and Fish and Wildlife Service) for providing working facilities and staff assistance; and to other organizations interested in marine mammalogy for special help on occasion. These organizations are mentioned by name at appropriate places in the text.

The writer will appreciate criticism of any phase of this work.

1

CHARACTERISTICS OF PINNIPEDS

Before entering a discussion of the evolution of pinnipeds it will be helpful to examine their characteristic biological features. It will be helpful also to note which features are unique (diagnostic) and which are shared with other mammals, especially the mammals which compose the superfamily Canoidea of the order Carnivora. The Canoidea include the dogs, raccoons, bears, weasels, and certain other less familiar mammals.

General Adaptations to the Environment

Most pinnipeds are marine, a few are lake-dwelling, all are remarkably adapted to aquatic life. While certain of the Carnivora (*e.g.*, the sea otter *Enhydra lutris*, the marine otter *Lutra felina*, and the polar bear *Thalarctos maritimus*) are specialized to obtain food from the sea, none has undergone a transformation in anatomy and behavior as complete as that of any pinniped.

Pinnipeds have retained in varying degrees an attachment to land. All must return to land (or ice) to give birth. Some (*e.g.*, the harbor seal) remain on land much of the time, others (*e.g.*, the northern fur seal) may spend six to eight months of the year at sea. Many of the subpolar species living at the edge of ice have developed migratory patterns in response to seasonal advantages in food and protection. It is believed that the northern fur seal, which travels the longest migratory path of any pinniped, may cover 10,000 km. in a year. A genus is occasionally represented by one population which is migratory and another which is resident (*e.g.*, the South Georgian as against the Uruguayan populations of the South American fur seal).

Outside of water the usual pinniped habitat is an island, offshore rock, sandbar, or ice floe; less often a broad mainland or extensive body of shore ice. Since seals are poorly equipped to defend themselves in combat with large land carnivores, they tend to frequent small, isolated breeding grounds which have sanctuary value. In the antarctic, entirely free of land mammals, certain seals breed on the shore ice and may at times travel long distances inland.

Pinnipeds are distinctly more sociable or gregarious than are land carnivores. Several species (*e.g.*, the Ross seal) are solitary in winter but gather in summer on breeding grounds. The size of a breeding ground population may vary from a few individuals to more than one million animals within a radius of 50 km. (*e.g.*, the northern fur seal). Advantages of the social trait in pinnipeds will be reviewed in Chapter 3.

An attempt has been made in table 1 to estimate the world populations of pinnipeds. The estimates are based on information of widely varying accuracy. At one extreme (*e.g.*, ringed seal of the Arctic Ocean) an estimate may represent little more than a rough calculation based on carrying capacity times total area of geographic range. At the other extreme (*e.g.*, Hawaiian monk seal) it may represent a reconnaissance of a small, isolated population or (northern fur seal) an intensive bionomic study.

Body Form, Covering, and Skeleton

Average body size in Pinnipedia is greater than in Carnivora. The smallest pinniped is the ringed seal: full-grown male, length 140 cm. or 55 in., weight 90 kg. or 199 lbs., skull condylobasal length 187 mm., skull weight 230 gm. For terminology of measurements see p. 135. The largest is the southern elephant seal: full-grown male, length 650 cm. or 256 in., weight about 3,629 kg. or 8,000 lbs., skull CBL 561 mm., skull weight 7 kg. Elephant seal, walrus, and Steller sea lion are each heavier than the largest land carnivore *Ursus arctos*: weight 771 kg. or 1,700 lbs. (Erdbrink, 1953, p. 367) and largest marine carnivore *Thalarctos maritimus*: weight 726 kg. or 1,600 lbs. (Anonymous, 1951). The large size of the pinniped has evolved mainly in response to a cold environment. There are in fact no small marine mammals of any kind. Cetaceans include the largest mammals ever known. The sea otter *Enhydra lutris* (39 kg. or 85 lbs.) is several times larger than any land otter of the genus *Lutra*; the extinct sea mink *Mustela macrodon* was "much larger than any of the living New England species" with which it was contemporaneous (Loomis, 1911, p. 227). Murie (1936, p. 342) found that at least 15 out of 25 walrus bacula excavated from St. Lawrence Island "exhibited more or less marked evidence of fracture." He suggested that evolution of body weight may be proceeding faster than evolution of the supporting framework.

Depending upon the social organization of the species to which it belongs, the male pinniped may be slightly smaller than the female (*e.g.*, monk seal, crabeater, leopard seal, Weddell seal, and perhaps all of the Monachinae), or about the same size (most of the Phocidae), or much larger (all Otariidae). In the northern fur seal the mean weight of the breeding male is 4.5 times that of the breeding female, a disparity greater than in any of the Carnivora. Laws (1956*c*) has compiled the lengths of the female in twelve pinnipeds.

Whether body size tends to vary in accordance with the so-called "Bergmann effect" is not clear. Scholander (1955) has in fact denied the existence of this effect, and he is probably right. While it is true that the Galapagos (equatorial) races of California sea lion and South American fur seal are smaller in skull than are comparable races at higher latitudes, one can easily point to contrary evidence. Thus the monk seal, the most tropical of all pinnipeds, is well above average size, while the ringed seal, distinctly polar, is the smallest.

Small but fully adult ringed seals are occasionally seen in groups of several hundred. Whether "dwarfism" here represents edaphic or genetic variation is open to question. The writer believes that, until more evidence is at hand, a formal name (*e.g.*, "*pygmaea*") should not be applied to members of such groups. Smirnov (1927, p. 19) concluded that "degenerated morphae are rather common in Phocinae." Freund (1933, p. 74) stated that the adult "Troldsael" (ringed seal) may be only 60 cm. in length, or half the usual length. Freuchen (1935, p. 196) regarded dwarfism as the result of limited food supply, which it probably is.

The near-maximum body size of a representative of each pinniped genus is given in the synoptic key (p. 135).

The pinniped body is streamlined—torpedo-shaped or spindle-shaped—though much less so than the body of the cetacean. The skull is somewhat telescoped, the supraoccipitals overlapping the parietals. The neck is thickened, more muscular, and yet more flexible than in most of the larger Carnivora. The thoracic cavity is elongated; head flattened (an advantage in diving); face shortened (except in the elephant seals and hooded seal); eyes situated well forward and rather close together, interorbital region narrow. The pelvis is more nearly parallel to the vertebral column than in Carnivora; tail very short and rudimentary, growing but little after birth, number of caudal vertebrae remaining unchanged. The external ears are reduced or absent; external genitalia and mammary teats at nearly all times withdrawn beneath the smooth surface of the body; the bases of the limbs to, or beyond, the elbows and knees deeply enclosed within the body; extremities flattened (hence the name "pinnipedia," the "wing-footed").

In the harbor seal (and probably all pinnipeds) the skin is adapted to a water environment (Montagna and Harrison, 1957). The epidermis is thick and tough; hairs are flattened and arranged in clusters; oil glands are large and numerous; the dermis is highly vascular. The significance of sweat glands and heavy pigmentation in the harbor seal skin is not obvious. Pinnipeds seldom, if ever, groom the pelage with the tongue.

The newborn pelage is often woolly, unlike that of the adult. Both the underhairs (sparse hairs as well as "fur" hairs) and the overhairs molt at regular intervals. Is the molt dictated by pituitary activity? Why, in a fairly uniform environment, does the seal molt at regular intervals? In the elephant seal the molt has been aptly described by Laws (1956*a*, p. 18) as "catastrophic," since wide tatters of epidermis and hair slough off together. The adult type of pelage in the pinniped is usually attained near the end of the first summer. In some seals (*e.g.*, harp seal and hooded seal) there are three well-marked coats: newborn, subadult, and adult. Though especially distinct in fur seals (Arctocephalinae) a two-layered arrangement of hair appears in all pinnipeds (except in the adult elephant seal and walrus, whose skin may be almost nude). The overhair fibers are coarse, stiff, and rooted deeply and singly; the underhair fibers are fine, soft, and rooted more superficially, in bundles (Troll-Obergfell, 1928, 1930; Bergersen, 1931).

Pelage color pattern in the adult pinniped is usually grayish or brownish; darker above than below; often spotted, flecked, or streaked; sharply banded in only two genera: ribbon seal *Histriophoca* and harp seal *Pagophilus*. A seal may appear silvery, brownish, or dark gray, depending upon the amount of water on the pelage, the amount of rookery stain, and the freshness of the pelage (since the last molt). Museum skins are invariably browned by oxidation of the natural oils in the skin. Since individual variation in the adult may be extreme (*e.g.*, in the harbor seal from uniform silvery gray through spotted to brownish black), one is led to suspect that color pattern in seals has little adaptive significance and is essentially a holdover from a land ancestor. Sex discrepancy in pattern is seen in the Otariidae and in the ribbon and harp seals. The mane of the adult male otariid is a secondary sex character and in cryptorchids is lacking entirely (Scheffer, 1951).

The evolutionary significance of color pattern is more clearly seen in the newborn than in the adult pinniped. The fetal or first pelage is shed in some species shortly before and in other species shortly after birth. So far as known, the first and second pelages are always unlike. Thus as a result of natural selection, the newborn animal may (depending on the species) be born in either one or the other pelage. A rough arrangement of newborn color patterns according to the probable significance of each is presented below:

1. Pupping centers on land

Newborn in dark first pelage: all sea lions and fur seals (Otariidae), monk seals, and elephant seals. These species have always lived in temperate regions?

Newborn in dark second pelage, first pelage light-colored and usually shed in the uterus: harbor seal. This species has invaded temperate regions "remembering" a past history of breeding on ice?

Newborn in light first pelage: grey seal. This species has bred on ice more recently than has the harbor seal?

2. Pupping centers on ice

Newborn in light first pelage: all seals not bracketed elsewhere. These have always lived in icy regions?

Newborn in dark second pelage: walrus. Why is the newborn walrus slaty gray rather than white? One can only guess that somewhere along its long and solitary course of development the walrus lost a whitish fetal coat, as it has now lost most of its adult hair.

Newborn in dark first pelage: Weddell and bearded seals. These have retained dark coats in memory, so to speak, of common ancestry with monk seals and elephant seals in temperate regions?

Pelage color in pinnipeds may change with the season. For example, the coat of the crabeater seal becomes almost white in late summer before onset of the annual molt (through the bleaching action of sun, water, and air?). These changes are generally inconspicuous; never as startling as the winter change in certain Carnivora such as arctic fox *Alopex lagopus* and ermine *Mustela erminea.*

Among northern fur seals on the Pribilof Islands about one in 100,000 born is completely albinistic (white pelage, pinkish iris and flippers). Partial albinism (*e.g.*, one eye pinkish and one black; one flipper pinkish and the others dark gray) is occasionally seen; also dilute coloration (chocolate over-all instead of glossy black). Albinism is reported in the southern fur seal and in the walrus and probably shows up from time to time in all pinniped species. Barnacles (*Lepas*) and several kinds of marine algae (*Ectocarpus, Enteromorpha, Erythrocladia*) occasionally fasten to the pelage of seals, especially in temperate waters. Completely nude specimens of the South American sea lion have been collected on the coast of Uruguay. Kumlien (1879, p. 60) remarked of the ringed seal: "There is sometimes caught a hairless variety . . . that the Eskimos call "okitook." I have seen one such skin. It had a few fine curly hairs scattered over it, but they were very different in texture from the ordinary hair." The present writer does not know that congenital atrichia has ever been observed in any pinniped.

The pinniped body is always enveloped in thick subcutaneous fat: the blubber. Bruce (1915, p. 3) found that about 113 kg. (250 lbs.) of blubber, or more than one-quarter of the weight of the animal, could be obtained from a Weddell seal. Blubber provides thermal insulation, reserve energy during lactation and during the fasts characteristic of most if not all pinnipeds, buoyancy, and padding. (By padding is meant the bulk necessary at certain points of the body to complete the streamlined profile.) Blubber is only moderately effective as an insulator. More important in maintaining the body heat of pinnipeds is the high basal metabolic rate; higher than in land mammals of similar size (Scholander *et al.*, 1950*a*, p. 234; 1950*b*, p. 266). Black *et al.* (1945), Thompson (1951), Bailey (1952), and Winter and Nunn (1953), have compiled data on the properties of pinniped blubber and carcass oils. Freund (1933, p. 66) reported that the iodine number of the fat in a fresh-water race of ringed seal was higher than in a salt-water race (values 195 to 202 as against 185). Is this a genetic or an edaphic character?

In cetaceans the blubber has similar functions. In the sea otter, which lacks blubber, a dense, waterproof, two-layered pelt provides insulation, buoyancy, and some streamline effect. It does not of course provide reserve energy. Thus mass mortality as observed from time to time among sea otter populations may perhaps be initiated and aggravated by lack of food during periods of stormy weather.

The number of vertebrae in pinnipeds is fairly constant and is diagnostic only for the walrus, which normally has 14 pairs of thoracic vertebrae (and pairs of ribs) and 6 lumbar; as against 15 and 5 for other pinnipeds. In 46 otariids and phocids for which vertebral counts were given by Flower (1884), Turner (1912), and King (1956), the range was: cervical 7, thoracic 15, lumbar 5, sacral 2–4, caudal 8–15, totaling 37–46. In all pinnipeds (as in Carnivora) the sternum normally has 9 sternebrae, rarely 8 or 10. The manubrium sterni is greatly elongated. As compared with those of most Carnivora the neck vertebrae are massive; the neck is muscular and often in motion. For example, a seal characteristically shakes a large fish into bits by rapid motions of the head, while a bear, using the fore paws, holds the fish to the ground and tears it. Interlocking processes of the pinniped vertebrae are reduced, permitting upward bending of the vertebral column. Some seals can touch the rump with the back of the head (see Mohr, 1956, fig. 3). "The cat," remarked Howell (1929, p. 23), "is looked upon as a rather limber mammal, but its vertebrae are far more securely interlocked than are those of the pinnipeds."

Locomotion

As will be explained in the key starting p. 135, the method of locomotion on land and in water varies with the superfamily.

The upper limb bones (humerus and femur) are extremely short. The extremity always has 5 digits (whereas that of the carnivore may have 4); digits partially concealed in weblike extensions of the flesh between digits; digits of fore limb decreasing in size from first to fifth (unlike the digits of any canoid); digits of hind limb unequal, marginal ones larger and stronger, though not always longer, than the middle three; digital bones in manus and pes numbering, as in carnivores, 3 - 4 - 4 - 4 - 4. Claw on each digit generally smaller than in carnivores, varying in length from a barely visible rudiment to about 56 mm. (on manus of male elephant seal; R. M. Laws, *in lit.*); front claws differing markedly in size from hind ones.

Angular tuberosities (ridges) on the long bones are reduced, since flotation does not require development of heavy muscles such as those needed for locomotion on land. The clavicle (collarbone) is lacking (rudimentary or lacking also in Carnivora); the fore limb is attached directly to muscles, with resulting flexibility; the humerus is massive and the bones of the fore arm are very broad; certain wrist bones—scaphoid, lunar, and centrale—are fused (as also in Sirenia); groove on acetabulum (hip socket) for ligamentum teres (from head of femur) lacking; groove on tibial facet of astragalus lacking (but present in the sea otter, suggesting longer history of pinnipeds as aquatic mammals); astragalus and calcaneum (heel bones) in the same plane.

Smith and Savage (1956, pp. 613, 615) have remarked that in the pelvic girdle the ilium is short and the ischiopubis is very long, giving more power at the expense of speed. As compared (percentage basis) with 18 other genera of mammals with assorted gaits, the harbor seal has the shortest femur and almost the longest pes; the tibia is of average length.

Detailed descriptions of pinniped musculature have been given by Miller (1888), Howell (1929, 1930), and Huber (1934).

The top swimming speed of a frightened northern fur seal has been estimated at 15 knots (17 m.p.h.) (Scheffer, 1950*a*, p. 30). A California sea lion (in captivity?) swimming underwater after food was clocked at 10.6 m.p.h. (Meinertzhagen, 1955, p. 115). A young hooded seal marked in April off Greenland was recaptured 13 days later and about 400 nautical miles away, having traveled at the rate of about 30 miles per day (1.43 statute m.p.h.) (Sivertsen, 1941, p. 61). Many kinds of pinnipeds have been seen leaping clear of the water, and it is likely that all seals, with the exception of the full-grown walrus and elephant seal, are able to do so. R. M. Laws writes (*in lit.*), "I have never seen *Mirounga* leap clear of the water and am sure that it is unable to reach the necessary exit speed." Unspecified antarctic seals were seen "rising 8 or 10 feet above the sea and covering distances of fully 20 feet in length" by Bruce (in Barrett-Hamilton, 1902, p. 4). A California sea lion about 8 years of age learned to hurdle a bar 216 cm. (85 in.) above the water (Anonymous, 1946*a*). Nansen (1925, p. 269) saw a hooded seal leap over a boat and upon an ice floe the height of a man above the water.

On land, the fastest pinniped is perhaps the crabeater seal. Lindsey (1938, pp. 458–59) described a race between a crabeater and a man sprinting at full speed (16 m.p.h.?) over a half-mile course on hard-packed snow. "The front flippers were used to strike the snow with strong backward strokes, but the principal motive force came from tremendous blows against the well-packed snow with the hind flippers. The latter were held palms together vertically behind the end of the body, like a caudal fin." Henry (1951, p. 23) remarked that antarctic seals tend to travel in left-turning circles on snow. Otto Geist reported that walruses may be forced by stormy weather to travel on snow 15 to 20 miles (24 to 32 km.) overland (in Murie, 1936, p. 342).

Diving will be discussed under respiration and circulation, p. 20.

Sensation

Brain weights in relation to body weights have been compiled by Freund (1933, p. 67) and Mohr (1952*b*, p. 58). In the full-grown South American sea lion the ratios are: 1 to 549 in the male, 1 to 308 in the female. "The brain of pinnipeds is distinguished by important development of the cortical motor apparatus, upon which the reflex movements depend; and the subcortical centers which receive and react to optic, acoustic, gravitational, and contact stimuli are considerably more developed than the cortical centers" (Frechkop, 1955, p. 303). The cerebral hemispheres are large and convoluted, with 3 or 4 concentric furrows around the fissure of Sylvius (Miller, 1952, fig. 180); whereas 2 in some Carnivora. The tentorium cerebelli is bony and very well developed.

The eyes are large, adapted to use at night and in deep or murky water; situated well forward. High-latitude seals (*e.g.*, Weddell seal and ringed seal) feed in total darkness for as much as four months at a time. The northern fur seal, certainly, and perhaps most seals, with the exception of the walrus, are nocturnal. The eyes of the walrus, a bottom-feeder, are small and piglike. Johnson (1893) described the eye of the harbor seal as very large, round, and of great convexity, with "myopia of 4 diopters . . . in the vertical meridian and 13 diopters . . . in the horizontal," obviously adapted to use under water. The iris of the harbor seal is narrow, vertical, and slitlike in daylight; large and circular under water. The iris in all pinnipeds may be described as dark-colored; at least it is never as yellow as in certain carnivores (lion and fox). The eye rests in a deep, protective cushion of fat; a lacrymal canal (tear duct) is lacking, though vestigial in otariids and the young walrus. Individuals totally blind but in good health have been reported (Rose, in Beebe, 1926; Bonnot, 1929, p. 53; Brimley, 1931; Scheffer and Slipp, 1944; Sorensen, 1950, p. 26). Freuchen (1935, p. 200) remarked of the ringed seal that "its sight is poor, whereas it hears well. In this it differs from the bearded seal . . . which has extremely keen eyesight but is hard of hearing." Variation in visual performance according to the species is rather great, yet, so far as studied, all seals can accommodate quickly from dim submarine light to brilliant sunshine. A thorough discussion of the pinniped eye has been given by Walls (1942, pp. 444–48); scattered through *Tabulae Biologicae* (1947, 1951) are numerical data on the anatomy and physiology of the eye.

The mystacial vibrissae (whiskers) are well developed (up to 400 in number in the walrus; up to 46 cm. or 18 in. in length in the Steller sea lion), richly supplied with blood vessels and nerves from the second branch of the trigeminus; infraorbital foramen correspondingly large. The superciliary vibrissae are fairly well developed in the Phocidae, feebly in the Otariidae; in both cases retrogression from a carnivore type seems to be under way. Mystacial vibrissae are remarkably developed in bottom-feeding walrus and bearded seal. Here the tremendous bouquet of bristles is thought to serve as a screen or filter for food organisms on the muddy bottom in addition to serving as a tactile organ. The bouquet in the male northern fur seal can be turned forward quickly during the moment of contact between his nose and that of the female. Courtship behavior?

Olfactory lobes and nerves are reduced; basioccipital and sphenoid bones are relatively large as compared with those of Carnivora. Studies of the northern fur seal on land suggest that the sense of smell is poorly developed as compared with that of, say, the sea otter. "In the Walrus and California Sea Lion, both carnivores but microsmatic [having poor sense of smell], there is only a very small and insignificant epiglottis, but with no disability in powers of swallowing" (Negus, 1949, p. 77).

The external ear cartilage is greatly reduced, leaving a small, stiff, pointed ear in the Otariidae and only a faint wrinkling of the surface of the skin in Odobenidae and Phocidae. The auditory opening is very small, 1 to 2 mm. in diameter. Ear muscles in the Otariidae are well developed and are able to close the opening, as in the river otter *Lutra*. Ear bones (tympanic bullae, auditory bullae) are massive, composed of endotympanic and tympanic bones which constitute the auditory canal.

The calls (voices) of pinnipeds are extremely varied. They range from the quiet notes of the Weddell seal (Bertram, 1940, p. 12) and the "high, chirping call similar to birds" of the Ross seal (Henry, 1951, p. 177) to the booming, staccato challenge of the northern elephant seal which can be heard for a distance of 2 km. The vocal apparatus of the California sea lion has been described by Kelemen and Hasskó (1931).

A good deal of practical information on the training of seals in captivity is available; little on basic psychology (Spindler and Bluhm, 1934; Hediger, 1955). Here is an inviting field.

Nutrition

The food of pinnipeds is exclusively flesh: fish, cephalopods, shellfish, macroplankton, sea birds, other seals, rarely small cetaceans, and carrion. While seaweed fragments are occasionally found in the stomach, these are probably ingested by accident. The diet is highly variable, according to the species of pinniped. Thus the crabeater feeds almost entirely on euphausians; the leopard seal on birds and seals; the harbor seal on many kinds of fishes, shellfishes, and squids. The northern fur seal eats at least 30 kinds of marine organisms (Taylor *et al.*, 1955, pp. 36, 56). Only one pinniped is known to feed on cetaceans; the walrus occasionally takes a narwhal *Monodon monoceros* or a beluga *Delphinapterus leucas.* (The writer has no firsthand information on this point, though he presumes that a walrus would more often attack a dead cetacean than a live one.) The food habits of the walrus are unique. Freuchen (1935, p. 244) has shown that the Eskimo actually "fish" for walrus, using strong lines baited with chunks of blubber.

In pinnipeds the mouth, jaws, teeth, and associated structures are developed for grasping and tearing, as opposed to chewing. Temporal and masseter muscles are reduced, with corresponding reduction of sagittal crest (fusion of temporal ridges) and postorbital processes. This tendency is partially countered in otariids and some phocids—especially the polygynous phocids—by the rival development of strong jaws for combat purpose.

As compared with those of most Carnivora the teeth of pinnipeds have become fewer and more uniform. (The walrus is an outstanding exception.) Carnassial cusps are not present and perhaps never were. (In Carnivora the carnassial teeth are certain postcanines—usually the 5th upper and 4th lower—equipped with lateral cusps for shearing flesh.) Reduction in the number of teeth from the basic carnivore formula is most frequent in the incisors and back molars; pinnipeds have only 1 or 2 lower incisors on each side; most of the Carnivora have 3. Except in the walrus, 1 canine and 5 postcanines are present and functional in each half of each dental arch in all pinnipeds. Except in the walrus, the postcanines are sharp-pointed, often conical, and never with more than 2 roots. Pinniped dentition in the normal full-grown adult is summarized below. In parentheses are indicated teeth of fairly common, though not regular, occurrence:

	Incisors	Canines	Postcanines	
Otariidae	1 - 2 - 3	1	1 - 2 - 3 - 4 - 5 -(6)-(7)	= 34–38
	0 - 2 - 3	1	1 - 2 - 3 - 4 - 5 - 0 - 0	
Odobenidae	0 -(2)- 3	1	1 - 2 - 3 -(4)- 0 - 0 - 0	= 18–24
	0 - 0 - 0	1	1 - 2 - 3 -(4)- 0 - 0 - 0	
Phocidae Phocinae	1 - 2 - 3	1	1 - 2 - 3 - 4 - 5 -(6)- 0	= 34–36
	0 - 2 - 3	1	1 - 2 - 3 - 4 - 5 - 0 - 0	
Phocidae Monachinae	0 - 2 - 3	1	1 - 2 - 3 - 4 - 5 -(6)- 0	= 32–34
	0 - 2 - 3	1	1 - 2 - 3 - 4 - 5 - 0 - 0	
Phocidae Cystophorinae	0 - 2 - 3	1	1 - 2 - 3 - 4 -(5)-(6)- 0	= 26–34
	0 - 0 - 3	1	1 - 2 - 3 - 4 -(5)-(6)- 0	

In the light of present-day information, the formulae for otariids and odobenids given by Winge (1941, p. 228) contain errors. The formulae for certain fossil species have been tabulated by Kellogg (1922, p. 84). The first four postcanines in each half of each dental arch in the adult pinniped are conventionally referred to as premolars. But since only the first three are preceded by deciduous teeth, perhaps the fourth "premolar" should be regarded as a molar. In the absence of general agreement on this point, the writer uses the inclusive term "postcanines" for all of the cheek teeth.

All pinnipeds are precocious; the deciduous teeth disappear before or soon after birth. Formulae for deciduous dentition as observed in the genera *Arctocephalus, Callorhinus, Odobenus, Phoca, Pusa, Pagophilus, Halichoerus, Erignathus, Monachus, Hydrurga, Leptonychotes, Cystophora,* and *Mirounga* are given below:

Otariidae	$i\frac{3}{2}$	$c\frac{1}{1}$	$pc\frac{3}{3} = 26$	(Rand, 1950*a*, p. 4; Chiasson, 1957, p. 311)
Odobenidae	$i\frac{3}{3}$	$c\frac{1}{1}$	$pc\frac{3}{3} = 28$	(Cobb, 1933, pp. 647–51)
Phocidae Phocinae	$i\frac{3}{2}$	$c\frac{1}{1}$	$pc\frac{3}{3} = 26$	(Allen, 1880, pp. 481–84; Kellogg, 1922, p. 84)
Phocidae Monachini	$i\frac{2}{2}$	$c\frac{1}{1}$	$pc\frac{3}{3} = 24$	(King, 1956, p. 236)
Phocidae Lobodontini	$i\frac{2}{2}$	$c\frac{1}{1}$	$pc\frac{3}{3} = 24$	(Bertram, 1940, p. 13; Brown, 1957, p. 22)
Phocidae Cystophorinae	$i\frac{2}{1}$	$c\frac{1}{1}$	$pc\frac{3}{3} = 22$	(Allen, 1880, pp. 481–84; Laws, 1953*b*, p. 16)

Cobb observed furthermore that the deciduous dentition of the walrus is inconstant; Brooks (1954, p. 27) gave a slightly different formula based on his interpretation of which teeth were deciduous and which were successional. Allen found in the bearded seal a peculiar "fourth, probably caducous, upper molar."

The tusk of the walrus is peculiar in that it carries enamel only at the tip for a short period after it has erupted. In the adult, the entire crown consists of dentin or ivory. Howell (1930, p. 84) believes that the evolution in length of the walrus tusk "if continued, threatens the extinction of this pinniped through constriction of the mouth to a degree which will hinder the ingress of food." A single tusk may attain a weight of 5.35 kg. (11.8 lbs.) and a length (curvilinear) of 100 cm. (39.5 in.)!

The pinniped mouth is simple and elongated; tongue notched at the tip (except in walrus), perhaps as an aid in sucking from a small teat; stomach simple, J-shaped, and aligned with the long axis of the body, perhaps in order to accommodate fish heads, fish vertebrae, and large chunks torn from prey; duodenal-jejunal flexure lacking; small intestine extremely long as compared with the intestine in most carnivores. Engle (1926) measured the whole intestine in an adult Steller sea lion at 80.5 m. (264 ft.), or about 38 times the body length. Laws (1953*b*, pp. 6, 15) measured the whole intestine in a male elephant seal at 201.8 m. (662 ft.), or 42 times the body length! Mohr (1952*b* and *in lit.*) found in various pinnipeds proportions ranging from 12 times to 22 times. (Length of the intestine in the dog is about 6 to 8 times the body length.) No entirely satisfactory explanation of the long gut in pinnipeds has been presented. Hediger (1955) termed the California sea lion in the zoo a "continuous feeder" as against "occasional feeders" like the great cats. Wheeler (1953, p. 254) has pointed out that "the outer layer of the ringed seal's intestines can be separated from the lining quite easily. This is also true of the bearded seal [but not of the harp seal]." (Biological meaning?)

The colon, cecum, and rectum are relatively short. (The cecum is lacking in all Canoidea except the Canidae.) The pinniped anus and vulva lie in a common furrow (as in cetaceans) and are closed by a common sphincter; anal glands seem to be lacking; liver moderately large and multilobed; gall bladder present (absent in porpoises, Delphinidae). The vitamin A content of the liver may at times be very high. Scheffer *et al.* (1950) found vitamin A in liver of the northern fur seal up to 413,000 "spec." units per gm. of liver oil. Nieman and Klein Obbink (1954, p. 80) have discussed the well-known fact that livers of bearded seal, polar bear, and beaked whale are often poisonous. However "the vitamin A–rich livers are more toxic than might be expected on the basis of their vitamin A content" alone.

Havinga (1933) found that charcoal moved through the alimentary tract of a harbor seal in captivity during a period of 6 to 14 hours after it was fed. From study of wild and captive animals he estimated that a seal weighing 100 kg. requires about 5 kg. fish a day, or 6,000 calories. Edward J. Johnson and staff of the Seattle Zoo found that charcoal passed in feces of a captive California sea lion at intervals of 12, 16, 18, and 38 hours after it was fed in herring. The cage was examined hourly. At 52 hours, no charcoal appeared in feces (the writer's data, unpublished). Scheffer (1950*b*) estimated that the northern fur seal eats one-fifteenth of its weight in a day; that is, a 100-kg. animal eats 6.7 kg.

Brooks (1954, p. 55) found 38.6 kg. (85 lbs.) of molluscs in the stomach of a walrus. A 600-pound male California sea lion in a New York zoo was fed 96 lbs. of fish a week; while a 70-pound female harbor seal was fed 28 lbs. (National Research Council, 1956, table 189). The total amount of food eaten in a year by members of one population—the northern fur seals of Alaska—has been estimated by the writer (1950*b*, p. 13) at 689,655 metric tons (760,000 short tons). Some if not all pinnipeds can fast for weeks or months: northern fur seal 64 days (K. W. Kenyon, *in lit.*); southern elephant seal in the wild over 2 months (Laws, 1956*a*, p. 69) and after capture 100 days or so (Hediger, 1955, p. 124); monk seal in captivity 4 months (King, 1956, p. 219).

The South American sea lion, California sea lion, northern fur seal, and presumably all other pinnipeds can regurgitate. However, one often finds the sharp remains of fish and shellfish lodged in the stomach of a seal, and Goodwin (1954) reported the death of a hooded seal from a large clam shell wedged in its throat. Huey (1942) examined a live but very thin northern fur seal whose stomach contained a mass of sea-bird feathers. Judging from the stage of molt, the feathers had been in the stomach for perhaps four months.

Stones are commonly found in the pinniped stomach. Some are quite surely ingested by accident; others by intent. Thus Laws (1956*a*, p. 17) has brought forth evidence that the southern elephant seal swallows stones before entering upon fast. Schneider (in Mohr, 1952*b*, p. 108) reported 31.5 kg. (69 lbs.) of stones in the stomach of a California sea lion in a zoo.

Seals defecate on land and in water (probably more often in water). Whereas land carnivores tend to defecate some distance away from the den, seals exhibit no instinctive behavior whatsoever toward sanitation.

Excretion

The surface of the kidney is grossly lobulated, giving the appearance of a compressed bunch of grapes (Harrison and Tomlinson, 1956, pl. 2). All pinnipeds tolerate salt water accidentally or deliberately ingested. Some species (*e.g.*, crabeater seal and walrus) feed mainly on invertebrate organisms, yet salt clearance does not seem to be difficult for them. Osmotic pressures of blood and urine in the harbor seal are only slightly greater than in man (Fetcher, 1939; Prosser, 1950, pp. 60–62). Perhaps it is unwise to draw a parallel, but Richter and Mosier (1954, p. 218) concluded that "Norway rats are able to handle very large amounts of salt and over long periods of time apparently without ill effects. . . On the highest concentrations they received an average of 7.5 gm/kg body weight of salt/day." (A harbor seal weighing 100 kg. could, at this rate, tolerate daily the amount of salt contained in about 20 liters of sea water.) Prosser (*op. cit.*) has concluded that "marine mammals living on fish [and invertebrates] can get ample water from their feed alone to keep their blood more dilute than the ocean . . . The urine of marine mammals can be more concentrated than sea water . . . The body should be able to extract some water from sea water and excrete salts . . . Many of the differences among animals from various habitats are differences in the degree of development of a capacity rather than in the specific kind of capacity."

During the breeding season the adult male northern fur seal may remain on land for over two months without taking water in any form by mouth. Irving *et al.* (1935) held 20 young harbor seals in captivity. "None of the seals were observed drinking sea water, but four seals which were shipped to Toronto in a warm express car drank *fresh* water greedily as soon as a clean supply was available." These authors calculated that, on a diet of fresh herring, seals obtained 1,121 g. water for each 1,000 calories metabolized, or 1,000 g. water in food plus 121 g. water of metabolism. Brown (1952, p. 982) stated that "the men at Heard Island constructed a large, mobile cage and drove a pregnant female leopard seal into it on September 12, 1951. Every three or four days the cage was dragged into the sea at low tide and left for the rest of the day. The seal would not feed during captivity but it was noticed that she drank large quantities of sea water upon being placed in the ocean." (She gave birth on 14 November.)

Renal calculi in Weddell seals were reported by Wilson (1907, p. 26) and Bertram (1940, p. 19).

Respiration and Circulation

The rhinarium (fleshy pad of the nose) is elastic and valvular; nostrils slitlike, normally closed, opened by voluntary effort; maxilloturbinale enormous, manifold, filling most of the nasal cavity; nasoturbinale reduced to a long slender bone near the ceiling of the nasal cavity.

The lungs are relatively longer and perhaps larger than those of land carnivores. The lungs of the ringed seal were found to be slightly heavier (percentage basis) than those of three land carnivores of similar body weight: cheetah, jaguar, and leopard (National Research Council, 1956, table 133). The diaphragm, instead of lying in a more or less vertical plane, slants from the backbone forward toward the sternum. The time pattern of breathing is very different than in land carnivores: periods of frequent breathing followed by long periods of rest. The nostrils return to closed position between each inspiration. Zeek (1951) discussed the peculiar 2-channeled trachea in the South American sea lion. Murphy (1913) pointed out that the trachea of the leopard seal has no rings but only short, transverse cartilaginous bars; the trachea "being a perfectly flat band 10 centimers in width" which can collapse to allow the passage of large-bodied penguins through the adjacent esophagus. The tracheal rings of the elephant seal are only four-fifths complete.

Maximum depth of diving is 60 to 80 fathoms (110 to 146 m.) in the Steller sea lion (Kenyon, 1952). Collett (1881, p. 387) reported that a young grey seal "was found caught by one of the hooks of a fishing line that was placed at a depth of between 70 and 80 fathoms [128 and 146 m.]." Bertram (1940, pp. 8–9) found Weddell seals in an ice-locked bay which, he believed, they may have reached by diving to a depth of 91.4 m. (300 ft.). Of the harp or saddleback seal Nansen (1925, p. 66) reported "it is said to be able to dive down to incredible depths. Thus off Rödöy in Helgeland several seals were captured in nets at a depth of 100 fathoms, and according to Robert Collett a saddleback was actually caught on a hook of a set-line 150 fathoms beneath the surface of the sea near Vardö." (Man has descended to 105 m. (344 ft.) in a flexible suit with normal air supply, and to 165 m. (540 ft.) in a flexible suit with special breathing mixture, according to Davis [1955, part 1, p. 26].) Maximum duration of diving is about 20 minutes in the ringed seal (Freuchen, 1935, p. 210), the grey seal (Backhouse, 1954), and the harbor seal (Scheffer and Slipp, 1944).

Müller (1940) has described and figured the heart in California sea lion, harbor seal, and elephant seal. The circulatory apparatus is specialized, although various aspects concerned with diving are shared by aquatic Carnivora and Cetacea. Adaptations are mainly responses to underwater pressure and apnea (cessation of breathing). The postcaval vein is greatly enlarged below the diaphragm into a singular "hepatic sinus." Here the blood flow is regulated by a muscular sphincter near the heart. The flippers are supplied with multichanneled arteriovenous bundles: the retia mirabilia, supposed to be heat-regulatory (Strays, 1956). During diving the metabolic rate falls, the cycle of heart beats falls to as low as one-tenth normal, body temperature falls; there is pronounced peripheral vasoconstriction, while the organs of prime need (heart and brain) continue to receive a normal supply of blood (Irving, 1942).

Bishop (1950, pp. 258–61) has concluded that "the major respiratory changes which make . . . diving possible are not the unusual storage of oxygen or increase in lung volume, but rather the more efficient utilization and availability of oxygen for the essential operating processes." In summary, he finds that critical factors in the diving mechanism are:

1. *Oxygen storing.* (*a*) Lung volume only slightly greater than in non-divers, though tidal air greater. (*b*) Blood volume probably greater. (In a harbor seal pup, the blood volume was found to be 117 ml. per kg. of body weight, as compared to 80 ml. in a human child and 70 ml. in a human adult [Harrison and Tomlinson, 1956, p. 223].) Oxygen capacity

of the blood hemoglobin higher than in man. (*c*) Muscle hemoglobin relatively abundant (note the dark red color of pinniped flesh). (*d*) Tissue fluids apparently the same as in nondivers.

2. *Other adaptations.* (*a*) Anaerobic glycolysis; oxygen debt builds up in muscle during dive; lactic acid, tolerated for 20 minutes or more, pours into the blood stream when the animal surfaces. (*b*) Carbon dioxide insensitivity. (*c*) Cardiovascular changes; a general slowing of circulation and shunting of blood away from peripheral muscles (as mentioned above).

The blood of the northern fur seal clots rapidly, in as little as 5 seconds under ordinary field conditions where dirt and impurities are present. (When blood is collected aseptically, clotting may be delayed for as long as 5 minutes.) Rapid clotting has survival value for a species whose members are subject to wounding by shark and killer whale; a species in which the males fight bloody battles and in which the umbilicus of the newborn snaps apart without being chewed by the mother.

The body temperature of adult northern fur seals at rest is about 37.7° C. (99.9° F.), of young seals 38.2° C. (100.8° F.). In the northern elephant seal and probably in all seals the temperature of the adult is lower than that of the young. Body temperature in pinnipeds is quite labile, rising or falling within a range of about 4° C. during ordinary activities. Mean blood temperatures (38.3, 37.2, and 38.5°) of three pinnipeds were found to be slightly higher than the mean blood temperature of man (36.9°) (National Research Council, 1956, table 311). Frechkop (1955, p. 296) stated that sebaceous and sudoriferous glands are well developed in pinnipeds; Bartholomew and Wilke (1956, p. 331) found that "both front and hind flippers [in northern fur seal] are abundantly supplied with large, well developed sweat glands." On a sunny day the northern fur seal cools itself by waving the hind flippers and opening the mouth; copious tears often stream down the creature's cheeks.

Seals of high latitudes (e.g., ringed seal and Weddell seal) maintain breathing holes in ice by tearing or sawing new ice with their teeth. The condition of the individual's teeth may spell the difference between life and death.

The phenomenon of sleep in seals has been widely observed and is probably universal. One can handle a sleeping fur-seal pup without waking it. Bartholomew (1952) lay prone on a sleeping elephant seal! "Typically sleeping elephant seals breathe more or less regularly about 30 times during a period of about 5 minutes, then spontaneously stop breathing for an interval which may last for more than eight minutes and which averages 5 minutes" (Bartholomew, 1955, p. 7). Northern fur seals habitually sleep at sea as well as on land.

Freuchen (1935, p. 198) saw ringed seals on the ice at all times of the year, even when the temperature dropped to —40° C. (—40° F.). No seal of course hibernates as do certain of the Carnivora (*e.g.*, bear, skunk, and raccoon) in cold climates.

Reproduction and Early Growth

The testes are either abdominal or scrotal and lie in a tunica vaginalis outside the body cavity but communicating with it by a small inguinal canal; vesicular glands and bulbo-urethral glands are lacking; these are lacking also in Cetacea. Prostate small; an os penis in all species, as in Canoidea. The os penis or baculum in the walrus may attain a length of 63 cm. (25 in.). There is (irregularly) a counterpart, the os clitoridis, in the female pinniped.

The ovaries lie in a broad, thin tentorium which communicates with the body cavity through minute openings; ovary smooth; follicles and corpora lutea barely visible on the surface; germinal epithelium (in six genera of otariids and phocids—perhaps in all (?) seals as well as many other mammals—dipping deeply, partition-like, into the body (tunica albuginea) of the ovary, forming crypts of various depths; ovaries normally alternating in function, annually (except in walrus, where biannually). In the elephant seal, at least, there is some evidence that ovulation may be induced by the act of copulation (Gibbney, 1957, p. 7). Uterus bicornuate, opening to the vagina by a single cervix, though partitioned for most of its length into two compartments. The 2-chambered uterus is a pinniped feature related to attachment of a single fetus in right and left sides in alternate years. The placenta is endotheliochorial (deciduous) and zonary (as also in Carnivora). The ischia (pubic bones) barely meet by a short symphysis (fibrous connection), are never fused, and in the female usually become widely separated; an advantage in giving birth to a large, precocious pup. Twin fetuses have never been found in dissection of more than 5,000 northern fur seals in gravid condition. Twin fetuses have been reported as a rare occurrence in the South American sea lion (Hamilton, 1939*b*, p. 135), South African fur seal (Rand, 1955, p. 732), harbor seal (Bertram, 1940, p. 27; Scheffer and Slipp, 1944, p. 405), harp seal (Bertram, *op. cit.*), and Weddell seal (Lindsey, 1937, p. 135; Bertram, *op. cit.*).

In spleen cells of the northern fur seal John L. Hamerton (*in lit.*, 1957) finds "a diploid chromosome number of almost certainly $2n = 36$. I would like to confirm this count on further material."

Slijper (1956) has summarized data on parturition in pinnipeds. The pup may be born in either breech or cephalic presentation. (One assumes that in breech presentation the pup does not start to breathe while its nose is still immersed in amniotic fluid.) The mother may tug at the placenta, cord, or membrane but rarely attempts to eat any of the afterbirth. All pinnipeds give birth on land or ice, though an occasional phocid or walrus may be born in water and yet survive. Tiny, almost hairless premature pups are occasionally found dead, rarely alive, on rookeries of the northern fur seal. The sex ratio in 2,172 late-term fetal or newborn northern fur seals was 51.7 percent males (Kenyon *et al.*, 1954, p. 19).

The mammary glands are thin, sheetlike, spread over the belly and sides; teats 2 or 4, abdominal, retracted when not in use. Harrison, Matthews, and Roberts (1952) have summarized the anatomical aspects of reproduction in pinnipeds and have given a 75-title bibliography.

Copulation takes place from a few days (northern fur seal) to four months (leopard seal) after parturition. Copulation is on land in all polygynous species, though between certain individuals (usually the younger ones) not yet hauled out on the rookeries it may take place in water. It is performed *more canem* in Otariidae and with the pair lying more or less side-by-side in the polygynous Phocidae (elephant seal, hooded seal?, and grey seal). Excellent sketches of coitus in the grey seal were given by Hewer (1957, p. 50). It takes place in water in monogamous species, including the great majority of pinnipeds; the exact posture is not known. Sivertsen (1941, pp. 72–79) discussed courtship display and mating in the harp seal in March. Polygyny is highly developed in all of the Otariidae. Depending upon the species, the average harem (or ratio of breeding females to males) varies from about 15:1 to 40:1. Polygyny is highly developed in the Phocidae only in the elephant seal (average harem about 20); moderately developed in the grey seal (average harem about 10). The evolution of polygynous behavior in pinnipeds is speculative. It is characteristic of certain seals which breed on land and of others which breed on ice; of seals which migrate and seals which do not; of seals which are abundant (successful) and seals which are rare. Polygyny in each group of pinnipeds may have risen out of a recurring local shortage of males. That is, in an ancient sociable stock the older males may have banished, year after year during the breeding season, an increasing proportion of the younger males. Today, the bachelor group situated outside of and distinct from the breeding group is a conspicuous feature of the social organization of polygynous species. Other concomitants of polygyny are sex disparity in body size, males fasting during the breeding season, and impregnating on land.

Delayed implantation is perhaps the rule in pinnipeds. It is known in the genera *Callorhinus, Arctocephalus, Mirounga, Pagophilus, Halichoerus*; probably in *Leptonychotes* and *Cystophora*; probably not in *Odobenus* (Harrison, Matthews, and Roberts, 1952; Backhouse and Hewer, 1956). In *Callorhinus,* for example, it extends over a period of 4 months, mid-July to mid-November.

The gestation period, from coitus, in pinnipeds is long in relation to maternal body size: 8 to 12 months, as compared to 3.5 months in the tiger and 7 months in the black bear (Asdell, 1946). Anatomical evidence suggests that toward the end of term the fetus is very active. It drinks and urinates; its colon becomes packed with ingested loose hairs and detritus, or meconium (Scheffer, 1945, p. 390; Scheffer and Slipp, 1944, p. 384). The newborn pup is always precocious, able to travel on land and to swim at birth, covered with protective pelage. Some species (*e.g.*, harbor seal and monk seal) may be obliged to swim on the day of birth; others (*e.g.*, all otariids) can swim but do not have blubber enough to provide insulation and buoyancy until an age of several weeks.

In Pinnipedia, as against Carnivora, the mother must often search for her pup among thousands of others in order to nurse it. Locating twin pups may be a time-consuming task. Perhaps the emaciated, solitary pup (*e.g.*, harbor seal) seen from time to time may represent one of twins, lost or abandoned. If true, mutations in the direction of twinning would tend to disappear from the blood line.

Hybrid young have resulted from intergeneric matings in captivity: grey seal male X ringed seal female; South African fur seal male X California sea lion female. (Fraser [1940] described certain odd dolphins which were probably intergeneric hybrids. Does intergeneric mating occasionally take place in the wild in a sexually excited aggregation of pinnipeds?)

As mentioned earlier, p. 17, the deciduous teeth of pinnipeds are rudimentary. Few are truly functional and all are lost before or shortly after birth. Growth during the nursing period is rapid; the mother's milk is rich—up to 53 percent fat in the grey seal (Amoroso and Matthews, 1951). Of 26 mammals for which chemical composition of the milk was given (National Research Council, 1956, table 50), "seal" with 42.0 percent fat had far richer milk than any other species. "Whale" was nearest, with 21.2 percent. In the Weddell seal, the young animal may have increased its neonatal weight by 50 percent at the end of the first week of life; by 100 percent at the end of the second (Bertram, 1940, p. 32). Growth in the southern elephant seal is even more astonishing (Laws, 1953*b*, p. 33). Here the average female weighs 101 pounds at birth. This weight is increased by 50 percent at the end of the first week, doubled at 11 days, trebeled at 17 days, and quadrupled at 21 days! The nursing period in the grey seal is from 2 to 3 weeks (Harrison *et al.*, 1952, p. 442); in the walrus as much as a year and a half (Chapskiy, 1936, p. 120). In pinnipeds, both the long gestation period and the long period of nursing on rich milk prepare the young animal to meet the sharp impact of weaning and the first winter at sea (Scheffer, 1950*d*).

Laws (1956*c*) has shown that, in the females of 12 species of pinnipeds, the body length at puberty is about 87 percent (81 to 92) of the full-grown length. Thus when she gives birth to her first pup the female is already well grown and prepared to meet the peculiarly heavy demands of the nursling.

Sexual maturity is attained rather late. In the northern fur seal some females are impregnated at beginning of their third year (Kenyon *et al.*, 1954, p. 40); in the walrus, at beginning of fifth year (Brooks, 1954, p. 50). In the harp seal "first ovulation takes place largely between the ages of five and eight, with a peak at six" (Fisher, 1956, p. 514). In the ringed seal, a few females may ovulate in the fifth year (McLaren, 1956). In the southern elephant seal "sexual maturity in the female is attained . . . at the age of two years" (Laws, 1956*a*, p. 46). Spermatogenesis may start in the northern fur seal in the third year; in the walrus in the fifth?; in the harp seal fourth; in the southern elephant seal fourth. For reasons that may be thought of as sociological, however, the males do not actually start to impregnate females for several years after the onset of spermatogenesis. (Compare another social species, man.)

Annual layers, the result of uneven deposition, may persist on the teeth, claws, or ear bones and thus become to the zoologist very useful indicators of the age of the seal. So far as studied, these layers are more prominent in high-latitude species, especially migratory ones, than in species of temperate waters (Scheffer, 1950*c*; Laws, 1952, 1953*c*; H. D. Fisher, 1954). However, the writer noted well-marked ridges on the teeth of a northern elephant seal from Isla de Guadalupe in latitude 29° N, and Kenyon (*in lit.*) has found them on teeth of the California sea lion.

In the process of developing adult habits of behavior, seals spend a good deal of time in play. Play is characteristic of the young; to a lesser extent of the adult. Eibl-Eibesfeldt (1955) has described the play of California sea lions on the remote Galapagos Islands. As do northern fur seals, these animals play with solid objects (seaweed and pebbles), engage in mock combat with their fellows, play at "King of the Castle," and ride the surf. Bartholomew (1952, p. 379) has found that play is a conspicuous element in the social behavior of elephant seals. "In this species, at least among adults and subadults, however, play appears to be confined entirely to the nonbreeding season."

Mortality

The average life of the northern fur seal (females, ages 1 year and older) is over 7 years (Kenyon *et al.*, 1954, p. 40). Of the harp seal "the maximum life span is well over 30 years . . . Animals of both sexes in their twenties are sexually active" (Fisher, 1956, p. 514). A male ringed seal in the wild was "a little over 40 years old" as indicated by tooth layers (McLaren, 1956). The record for longevity seems to be held by a grey seal which died at age 41 or 42 years in a zoo (Matheson, 1950).

Seals, especially the young, may be crushed by rolling stones and trod upon by adult bulls as much as 50 times the weight of the pup; they may fall into pits and drown or starve; they may be submerged and beaten by storm-driven waves and ice floes; they may die of exposure at sea during prolonged storms; they may be smothered under sea ice; they may die as a result of intraspecific combat or misdirected sex activity on the part of the adult male. Beebe (1926, p. 144) saw a sea lion scalded to death by a submarine flow of lava. Not infrequently a pinniped dies in the act of parturition. Cumulative natural mortality in both sexes of the northern fur seal is estimated at 72 percent up to the beginning of the second year of life (Kenyon *et al.*, 1954, tables 14–15).

The marine predators of seals include large sharks in temperate and warm waters, the killer whale *Orcinus orca* (especially in colder waters), the leopard seal, and walrus. Mortality among pinnipeds as a result of predation probably diminishes sharply after the young animals have learned to swim easily and to avoid pursuit. Writing of *Zalophus* in the Galapagos Islands, Heller (1904, pp. 244–45) stated that "the sharks, chiefly the genera *Carcharhinus* and *Galeocerdo*, are the worst enemies the seals have to contend with. Their depredations are confined largely to the pups, though the latter genus is a serious menace even to the adults. While the crew were engaged in collecting shark fins, we had an opportunity of dissecting a large number of sharks, and found that a majority of those caught contained the remains of seals, chiefly pups." Eschricht (1866) dissected at Jutland a male killer whale whose stomach was distended to dimensions of 4.5 by 6 feet with fragmentary carcasses of 13 seals and 13 common porpoises in various stages of digestion! Admitting the possibility of error in count of individuals, the total number of prey items must have been impressive. The walrus is known to kill and eat ringed seals and bearded seals (Brooks, 1954, p. 57). (The walrus seems to be poorly equipped to take this kind of prey.)

In arctic seas the polar bear *Thalarctos* is an important predator of seals and walrus. Soper (1928, p. 35) noted that "during March and April the foxes [*Alopex lagopus*] captured very many of the helpless young of the ringed seal in their snow dens on the ice." Stefansson (1943, p. 354) reported the killing of a seal by a wolf as being a rare event. Brandenburg (1938, p. 47) found remains of young sea lions in a cave in Patagonia where a mountain lion (*Felis concolor*) was later trapped, and he saw lion tracks at a rookery on the beach. The writer has no information on predators of seals of frozen inland waters such as Lake Baikal.

In Alaska the bald eagle *Haliaeetus leucocephalus* occasionally kills young harbor seals. In antarctic regions the giant petrel *Macronectes giganteus* attacks weak elephant seal pups (Laws, 1956*a*, p. 23). Vampire bats (*Desmodus*) are said to attack sea lions in Chilean caves (Mann Fischer, 1955, p. 17), surely a most unusual relationship.

Weddell seals of the antarctic, monk seals of the Hawaiian Islands, and California sea lions of the Galapagos Islands are said to be very tame.

Is this because they have never suffered predation on land? The men of the *Scotia* occasionally diverted themselves by riding on the backs of Weddell seals (Hayes, 1928, p. 105).

Little is known, unfortunately, about the diseases of pinnipeds in the wild. (Mohr, 1952*b*, pp. 103–13, has reviewed the principal ailments of seals in captivity.) No epizootic has been observed under circumstances which permitted identification of a primary pathogen. Fleming (1828, vol. 1, p. 17) wrote that "about fifty years ago, multitudes of carcasses [grey seal?] were cast ashore in every bay in the north of Scotland, Orkney and Zetland, and numbers were found at sea in a sickly state." Morrell (1832, p. 290) was puzzled to find "not less than half a million" fur seals dead on Possession Island, South-West Africa. In autumn 1947, numerous dead sea lions were seen on the beaches of California and "health authorities . . . diagnosed the cause of death to be streptococcal pneumonia" (Anonymous, 1948). Shortly after this report the present writer attempted without success to obtain factual information upon it. Laws and Taylor (1957) have described the finding of several hundred carcasses of crabeater seals and have supposed that disease spread through a small detachment. Scheffer (1950*a*, p. 26) reported 13 carcasses and, on another occasion, 40 carcasses of the northern fur seal found dead from unknown causes on a small Alaskan island. Does the toxin produced by "red tide" organisms (dinoflagellates) have a serious effect on seals? Sealers in both Northern and Southern Hemispheres are subject to a serious infection known as "seal finger," its etiology unknown. The agent may be a virus (Skinner, 1957).

Freuchen (1935, p. 219) suggested that ringed seals died simultaneously from "dangerous spines of the armed bullhead (*Agonus cataphractus*)." Skin disease (sarcoptic mange? ringworm? physiological decline?) is commonly seen among northern fur seals but has not been studied. Gwynn (1953*a*, p. 23) stated that "Chittleborough and Ealey noted several [leopard seals] with extensive skin disease; this affliction was also noted by the writer, and in some cases was extremely severe . . . One seal seen by the writer had lost all the hair off the greater part of its body, and appeared weak and ill."

Seal parasites are common. While certain species may be abundant in crowded populations (*e.g.*, hookworm, *Uncinaria lucasi*, in the northern fur seals of the Pribilof Islands) they cannot be said with assurance to be primary causes of mortality. Mohr (1952*b*, pp. 104–5) gave a list of 50 parasites of seals (phocids and walrus) of European waters. These included trematodes (flukes), cestodes (tapeworms), nematodes (roundworms of stomach, lungs, and heart), acanthocephala (thorny-headed worms), arachnoidea (nasal mites), and lice. (A tick *Dermacentor* reported from walrus was surely a sea-bird parasite temporarily associated with a marine mammal.) Rausch *et al.* (1956) gave positive records of *Trichinella spiralis* in walrus, ringed seal, and bearded seal. How these pinnipeds became infected was not known. Margolis (1954, 1956) and Duncan (1956), among others, have compiled lists of seal parasites.

The effect of water and air temperatures upon incidence of disease and parasitism in seals has been little studied. O. Wilford Olsen has reported (*in lit.*) that larvae of the fur-seal hookworm are found hibernating in soil where air temperatures drop to —32° C. (—26° F.). Rose (in Beebe, 1926, pp. 153–56) observed conjunctivitis among nearly all members, young and old, of a small detachment of sea lions on the equator.

2

EVOLUTION OF THE ORDER PINNIPEDIA AND ITS FAMILIES

The Order

With respect to its ability to satisfy the basic requirements of the mammal for food, shelter, and breeding space, the ocean is a limited province. Only three groups have been able to exploit it: the Cetacea (completely pelagic) and the Sirenia and Pinnipedia (littoral). Furthermore, of the Sirenia only the dugongs are truly marine; the manatees are estuarine. The now extinct Steller sea cow, when discovered in the eighteenth century, was clinging precariously to life on two small islands in the Bering Sea. The Carnivora are trying on a small scale to exploit the ocean through three representatives: first, the sea otter, a specialized and not too successful mammal which has remained among the reefs of the North Pacific; second, the polar bear, a mammal which is hardly a marine creature and is able in fact to interbreed with the brown bear; and third, the marine or Cape Horn otter *Lutra felina,* which differs but little from the common river otter.

Kellogg (1922) gave an excellent and instructive résumé of theories on the origins of the Pinnipedia. He included a check list of fossil forms with their geographic and geologic time ranges. (See also his 1936 comparisons with primitive cetaceans.) Howell (1929, 1930) surveyed the evidence, from studies of comparative anatomy, on blood lines of the Pinnipedia. Newell (1947) and Hopkins (1949) discussed the origins of the Pinnipedia in the light of information on their parasites. Simpson (1945), Viret (1955), and Downs (1956) commented on various theories of origin.

The oldest known pinniped remains are from the Miocene. (Several bone and tooth fragments from older beds are of uncertain relationship.) "The pinnipeds were well specialized in the Miocene for a pelagic life and the distribution of the fossil forms corresponds very well, in most respects, with the present distribution of their living representatives" (Kellogg, 1922, p. 93). The oldest Miocene beds have an age of perhaps 35 million years. It is generally agreed that search must be made in horizons two or three times older (Eocene or even upper Cretaceous) for the progenitors of the pinnipeds.

Downs (1956, pp. 130-31) has commented on the singular absence of pinniped remains in pre-Miocene rocks containing cetacean bones. "These rocks surely could have contained otariids if they were amply represented in the marine waters . . . The original population of land carnivores from which the pinnipeds were probably derived must have been small or rare, geographically restricted, genetically variable, pre-adaptively suited, and receptive to effective selection in order for them to cross the 'threshold' and enter the new, aquatic zone. The possibility that the original transitional forms were members of small and rare populations (fresh or marine water inhabitants) may explain the lack of fossil evidence."

While the progenitors of the pinnipeds are completely unknown, three alternative reconstructions have been proposed:

First, origin from as yet undifferentiated insectivore-creodont stock in the late Cretaceous at about the time the Cetacea and Sirenia were splitting off. Although modern pinnipeds, retaining as they do hind limbs, typical eutherian dentition, and hairy covering, are less modified than cetaceans and sirenians, they may be equally ancient, stemming from the Cretaceous. That is, the dependence of pinnipeds upon land at the breeding season may have retarded, so to speak, the development among them of extreme modifications for aquatic life such as are manifested by cetaceans and sirenians.

Second, origin from creodonts in early Eocene along a line distinct from that followed by fissipeds. "Common derivation with the land carnivores is now universally admitted," concluded Simpson (1945, p. 232), and so regarded the Pinnipedia as a suborder of the Carnivora. The present writer is inclined to favor this theory of origin but feels that the Pinnipedia deserve recognition as a separate order.

Third, origin from fissiped stock in late Eocene. Modern pinnipeds most closely resemble members of the dog-raccoon-bear-weasel group (superfamily Canoidea of Simpson; Arctoidea of Winge); less closely the civet-mongoose-hyena-cat group (superfamily Feloidea). Mivart (1885) pointed out resemblances of otariids and phocids to bears and otters, respectively. Kellogg has warned, however, that "we must look for the ancestral otarids among the Carnivora long before there were any true bears or wolves" (1922, p. 100). In short, similarities between bears and "sea-bears," wolves and "sea-wolves" represent convergence.

As the fossil record has given few clues, so has comparative anatomy of modern forms provided but little evidence on the origin of the pinnipeds. Many writers (especially Howell) have pointed out resemblances between this organ and that one in pinnipeds and in carnivores. Eadie and Kirk (1952) showed that the concentrations of N and K cations in blood of the elephant seal are more like those in the cat and dog than in other mammals, suggesting a "common physiological inheritance."

Friant (1956, p. 267) concluded that the "Pinnipedia are doubtless arctoid Carnivora adapted to aquatic life." Leone and Wiens (1956, p. 22) concluded that "the classification of the Pinnipedia as a separate suborder is not justified on the basis of the serological data which places them within the Canoidea." That is, from observation of the reaction of fur-seal serum against the sera of various land carnivores, these authors concluded that pinnipeds as a group deserve no higher rank than that of a family.

So remote, however, are the ancestors of the Pinnipedia that little evidence upon them can be expected from studies of the comparative anatomy and physiology of modern forms. Exploration of marine or even fresh-water pre-Miocene beds would seem to offer greatest promise of reward.

The ancestral home of the pinnipeds may have been in the Northern Hemisphere where 2 of the 3 families and 12 of the 20 genera are represented today.

The Families

It cannot be shown that any of the three families Otariidae, Odobenidae, or Phocidae is ancestral to another. Each has developed peculiar features adapting to a way of life. Regard, for example, the dense skeleton and unique crushing, not lacerating, dentition of the walrus. Such features have destroyed or overridden many of the anatomical structures that once memorialized the history of evolution in the order. In general, though, members of the family Otariidae most closely resemble land mammals; especially in walking gait, copulation *more canem*, dentition, external ears, hairy or furry pelage, scrotal testes, and greater average number of mammary teats (four). The Odobenidae resemble the Otariidae in the fundamental arrangement of the limbs. Some distance apart from these two families are the Phocidae with body and limbs more specialized for pelagic life. The gap between the superfamilies Otarioidea and Phocoidea has suggested to some students dual origin of the Pinnipedia. A gap, they say, may have started to widen on land among the Creodonta (?), with members of each subdivision later and independently taking to sea (*cf.* Huber, 1934, p. 113). Assuming dual origin, Howell (1929, p. 139) stated that "the present contribution sheds little or no light upon the question of whether the Otariidae or the Phocidae is the 'older' family. The evidence is conflicting and the proper weight to accord details of variation is, and probably always will be, a moot question. It is felt, however, that this evidence points to the probability that the Otariidae although not necessarily the better (or as well) equipped for an aquatic existence, have perhaps departed widely from a typical terrestrial condition in more numerous and profound respects than the Phocidae." Howell described as advanced features of the otariid "the 'telescoping' of its skull, cartilaginous extensions of the digits, the greater tendency toward flattening exhibited by the pedal phalanges . . . and possibly by the development of the nails" (p. 138). Most of these changes, however, seem to be superficial, and the present writer is inclined to disagree with Howell; to regard the Phocidae as more specialized—certainly far more diverse, numerous, and widely distributed—than the Otariidae. The Odobenidae lie somewhere in between and nearer the Otariidae.

Hopkins (1949) noted that the lice (Anoplura) of pinnipeds can be placed in a unique systematic group. Furthermore "the fact that all three families of seals are infested with closely related parasites is an exceedingly strong indication that the infestation is primary, especially in view of the fact that the seals can have had practically no contact with other louse-infested mammals since they took to aquatic life" (p. 546). He concluded, first, that seals from the time they were creodonts, so to speak, have had lice and, second, that the Pinnipedia are of monophyletic origin.

Newell showed, significantly, that certain nasal mites peculiar to pinnipeds can be arranged in two genera: *Orthohalarachne* parasitic in otariids and walrus only, and *Halarachne* parasitic in phocids only.

1. *Otariidae.* The Otariidae may have originated in the North Pacific. No otariid has been recorded from the North Atlantic and no sea lion from the eastern South Atlantic. "Most writers have assumed that the distribution of the Otariidae in the past was much the same as it is today. Since three distinct forms with otarid relationships [*Allodesmus, Desmatophoca,* and an unnamed fossil from the Vaquero formation] are known from the Pacific coast during the Miocene, one is led to believe that they must have had their origin somewhere in the North Pacific Ocean" (Kellogg, 1922, p. 66). The South Pacific is equally a possibility. However, from the fact that sea lions are unknown from the polar quadrant centering in South Africa, one supposes that these members of the Otariidae have not yet had time to colonize all of the cool-water islands surrounding Antarctica. One supposes, in short, that the Otariidae date more recently from the Southern Hemisphere than from the Northern. (It is perhaps more than coincidence that the sea otter, representing an experiment in the adjustment of a carnivore to marine life, inhabits only the North Pacific. Did the protected, food-rich, kelp reefs of the North Pacific serve as a launching platform for both otariid and sea otter stocks?)

The oldest known member of the family is *Allodesmus kernensis* Kellogg (1922, p. 26), described as a new genus and new species on the basis of a right mandibular ramus found in Miocene (Temblor) beds near

Bakersfield, California (plate 1). *Allodesmus* was large and heavy, suggesting kinship with the walrus. It had a sixth lower postcanine, suggesting derivation from a land carnivore. No modern adult pinniped has a sixth lower postcanine (exceptionally *Halichoerus* and *Mirounga*). *Allodesmus*, however, is not greatly unlike *Eumetopias* now living along the California coast. Downs (1953) estimated the age of the Temblor beds as middle to late Miocene. He found no apparent relationship between *Allodesmus* and the bears (Ursidae).

Another ancient otariid, *Arctocephalus fischeri* (Gervais and Ameghino, 1880, p. 222), was described from a left mandible found in Miocene (Piso Patagonico) beds, Province of Paraná, Argentina. The specimen closely resembles *Arctocephalus australis*, a living form (Kellogg, 1922, p. 59).

2. *Odobenidae.* With regard to primitive distribution of the walruses, Kellogg (p. 51) stated "it seems probable that they had their origin in the North Pacific and that during Oligocene time they migrated to the Atlantic by way of the sea which then separated North and South America." Quite certainly as a result of lack of exploration, fossil odobenids are unknown from Europe below middle Pliocene and from North Pacific shores below Pleistocene levels.

The oldest known member of the family is *Prorosmarus alleni* Berry and Gregory (1906, p. 447), described as a new genus and species on the basis of a left mandibular ramus from the upper Miocene (Yorktown), near Yorktown, Virginia. *Prorosmarus* resembles the otariids in having two pairs of lower incisors, as against none in adult *Odobenus;* the canine in primitive position and caniniform, as against pushed backward and molariform in adult *Odobenus;* and in other respects. "It seems reasonable to assume from what is known of *Prorosmarus* that it possessed a skull somewhat similar to that of an otarid, with upper canines or tusks much enlarged" (Kellogg, 1922, p. 53).

3. *Phocidae.* Eight phocid genera are northern while only five are southern (counting *Mirounga* as southern). The northern phocids are more diverse—taxonomically more complex—than the southern ones. Thus it is generally supposed that the family Phocidae, as well as the other two families, originated in the Northern Hemisphere.

The oldest known member of the Phocidae is *Leptophoca lenis* True (1906, p. 836), described as a new genus and species on the basis of certain skeletal bones (no skull) from the upper Miocene (Tortonian) of Maryland. A seal tentatively identified as a phocid by Kellogg (1922, p. 70), from lower Miocene beds at Lompoc, California, proved on later examination to be an otariid, *Pithanotaria starri* Kellogg (1925, p. 74). The phocid line may be very ancient. Kellogg (1922, pp. 66-91) has pointed out that some of the earliest known phocids appear to belong to the same subfamilies as do recent forms. A tooth of "*Lobodon vetus*" (Leidy) 1853, found in upper Cretaceous sands of New Jersey, has been questioned. Phocid remains are fairly well known from North Atlantic and North Pacific beds deposited since upper Miocene time.

4. *Fossil families.* In addition to the Otariidae, Odobenidae, and Phocidae, three fossil families were listed by Simpson (1945, p. 121). The first two, Desmatophocidae Hay (1930, p. 557) and Allodesmidae Kellogg (1931, p. 227), were regarded by Simpson as synonyms of Otariidae. He was probably right in his interpretation, for the type genera *Desmatophoca* and *Allodesmus* closely resemble, and are quite certainly ancestral to, otariids living today near the fossil-type localities on the Pacific coast. The third family, Semantoridae Orlov (1931, p. 69), listed as valid by Simpson, was later knocked out by Thenius (1949). It was based on remains of the hind part of the skeleton of an animal, probably an otter, from the lower Pleistocene. It was far too late an arrival on the geologic scene to have served as a "missing link."

3

EVOLUTION OF THE GENERA, SPECIES, AND SUBSPECIES

Among zoologists who have attempted to trace the evolution of modern pinnipeds are Balkwill (1888), Trouessart (1881, 1922), Grevé (1896), Sclater (1897, 1898, and critical reviews of his work by Baur, Gill, and Ortmann), Kellogg (1922 and later), von Boetticher (1934), Croix (1937), Ekman (1953), and Downs (1956). The present writer suggests that existing genera, species, and subspecies of pinnipeds are the result of evolution in four stages: invasion, geographical isolation (usually with ecological isolation), dispersion, and stabilization.

Invasion

The early pinnipeds moved out in narrow file along the shores of continents, from island to island and later to polar ice fields. They ranged essentially in one dimension rather than two. Even today all pinnipeds attach themselves to a land front or an ice front during the reproductive season. Since its geographical ends are farther apart, a population extending itself along a thin line provides more opportunity for the development of clinal distinctions than does one expanding fanwise or radially. The curious fact that 20 genera of pinnipeds have developed in what appears to be a rather monotonous environment—the sea—is related quite surely to the fact that the primitive stocks followed separate, distinct, linear, branching pathways. Some of the pathways have persisted down to the present day as long lines of distribution. That of the South American sea lion is V-shaped and 10,000 km. in extent; that of the harbor seal is broken-circular with four radial extensions, 25,000 km. in total extent.

The rate of evolution was more rapid at the spearhead of the advancing line, where immigration was from one direction only, than near the middle. It is suggested that the primitive populations were small, composed of family groups, not very sociable. The members did not wander far, had no well-developed homing instinct, were not migratory, and did not need to rendezvous in special places in order to find mates. Polygyny had not developed in any of the pinniped stocks. The animals were perhaps smaller than most recent pinnipeds (Cope's rule), had less fat, lived in more temperate waters, and perhaps made crude dens in beach grasses and among boulders. The advancing pinnipeds met, from sea birds and cetaceans, little competition for food. They met no competition for breeding room, nor do they often today. Pinnipeds early lost the habit of feeding on beach organisms such as crabs, mussels, periwinkles, and blennies. Without exception and in spite of widely diverse food habits, all modern pinnipeds feed under water.

Geographical Isolation

Discontinuities appeared in the line from time to time. Rarely did these follow the rise of important physical barriers, for the marine environment was already old and stable when the protopinnipeds arrived on the scene. The sea has in fact seldom been exposed to influences as profound as those which disturb the face of the land from time to time: fire, flood, drought, and dust storm. As local shore lines and islands rose and fell and glacial barriers came and went, pinnipeds moved back and forth in order to maintain favorable breeding grounds along the edge of the sea. (On land, many of the contemporary Carnivora were being forced into extinction.)

Many—and perhaps most—discontinuities in the line were present from the start of the invasion. There was, for example, at no time a continuous line of distribution of primitive otariids between Australia–New Zealand and adjacent continents (assuming that otariids did not then, as they do not now, breed along ice fronts). Island-hopping took place from time to time during periods which must be reckoned in millions of years. Australia and New Zealand were reached quite early by a sea lion stock, for they now harbor a distinctive genus *Neophoca.* The Hawaiian Islands were only recently colonized, for the monk seal *Monachus* here is very like—and perhaps only subspecifically distinct from—the one in the Caribbean. On the basis of present fossil information, South Africa has not yet been reached by a sea lion.

Certain other discontinuities in the line of advance appeared rather in the fashion that droplets separate by surface tension from a line of water drying on a table top. Suppose, for example, that an ancestral fur seal less specialized than any living today pushed southward through tropical waters. The northern populations and the southern populations then independently developed certain attributes of modern seals, for example, a thick blubber layer. For reasons explained on p. 11 the blubber fat has important survival value. But to populations approaching the tropics the blubber layer was an embarrassment, and finally a lethal accoutrement, especially to young seals. The line of distribution snapped apart and today one finds the breeding metropolis of *Callorhinus* in subarctic waters and that of *Arctocephalus* in subantarctic waters.

(Yet the same species of nasal mites, reflecting a common ancestry, are found in both [Till, 1954].) In similar fashion it may be supposed that other north-and-south pairs of genera were separated: *Eumetopias* and *Otaria*, *Zalophus* and *Neophoca*. Huxley (1954, p. 9) has aptly stated that "specialization . . . forces organisms into a deepening evolutionary groove out of which it is increasingly impossible for them to climb."

Thus as each generic stock became isolated it began to experience transformation as a result of influences in its own physical and biotic environment. It became molded into an ecological niche. Each of the stocks retained the characters of the order and family but began to deviate at lower levels. For example, somewhere near Antarctica, though not at one place (or at least not at the same time and place), the four generic stocks of the Lobodontini began to separate. Today, *Lobodon* is almost exclusively a feeder on euphausians, *Ommatophoca* a feeder on soft-bodied cephalopods and fishes of the outer pack ice, *Leptonychotes* on fishes and cephalopods of the shore ice. *Hydrurga* is largely a predator on sea birds and seals. In arctic seas no ecological counterpart of *Hydrurga* has chanced to evolve (perhaps because of the absence of penguin-like birds).

The partitioning of the Lobodontini shows beautifully the process of evolution in two planes: ecological or vertical as well as geographical or horizontal. In the formation of certain other genera, for example *Eumetopias* and *Otaria*, geographical isolation has played an important part (as indeed it must have played in the branching of all phyletic stems) but ecological isolation very little. The habits of the two sea lions *Eumetopias* and *Otaria* seem to be similar. (Is it presumptuous to suggest that, if pups of *Eumetopias* were placed on breeding grounds of *Otaria* in comparable climatic zones, the northern animals would be accepted by the southern group?) The ribbon seal *Histriophoca* and the harp seal *Pagophilus* also represent paired genera geographically, but not ecologically, far apart.

Dispersion

During the evolution of the genera, ancestral seals moved about from one sea to another, in increasing numbers and for longer distances as they developed the attributes of pelagic creatures. At an early period a nomadic detachment would have been accepted by, and genetically absorbed by, a resident colony. At a period much later in geologic time a similar detachment would not have been accepted (or in a manner of speaking would not have sought to be accepted) and would have colonized alongside the "natives," provided competition were not too great. Today one can see as a result of dispersion and intermingling members of two genera breeding side by side in complete reproductive isolation (*e.g.*, *Eumetopias* and *Callorhinus*, *Leptonychotes* and *Lobodon*). And one can see (with luck) members of as many as six genera feeding at the same place (*e.g., Odobenus, Phoca, Pusa, Pagophilus, Erignathus,* and *Cystophora* along southern Greenland; or *Zalophus, Eumetopias, Callorhinus, Arctocephalus, Mirounga,* and *Phoca* along southern California). (See table 2.)

TABLE 2. BREEDING RANGES OF PINNIPEDS BY GENUS AND OCEAN*

Genus	Atl.-Arct.	Pac.-Arct.	Temp. N. Atl.	Temp. N. Pac.	Temp. S. Atl.	Temp. S. Pac.	Ant.-arct.	Note
Otaria					●	●		
Eumetopias				●				
Zalophus				●				
Neophoca						●		
Arctocephalus				O	●	●	O	†
Callorhinus				●				
Odobenus	●	●						
Phoca			●	●				
Pusa	●	●	O					‡
Histriophoca		●						
Pagophilus	●							
Halichoerus			●					
Erignathus	●	●						
Monachus			●	●				
Lobodon							●	
Ommatophoca							●	
Hydrurga							●	
Leptonychotes							●	
Cystophora	●							
Mirounga				O	O	O	●	

* ● = principal breeding range and O = limital breeding range in Atlantic-Arctic, Pacific-Arctic, temperate North Atlantic, temperate North Pacific, temperate South Atlantic, temperate South Pacific, and Antarctic (= Southern) oceans.
† And extreme southern Indian Ocean.
‡ And inland waters of northern Eurasia.

As they dispersed, a few pinnipeds were eventually able to penetrate barriers which had been important at an earlier time in holding the generic stocks apart. For example, certain members of the genera *Zalophus, Otaria,* and *Arctocephalus,* and especially all members of *Monachus,* now live in small numbers in subtropical latitudes, apparently able to survive by taking advantage of cool caves, rock shadows, mud wallows, sand baths, and water baths. Other barriers are yet unbroken. Thus the southern elephant seals and southern fur seals have not been able to pass through the tropics and into the North Atlantic; and the Bering Sea otariids have not been able to push either eastward or westward through the Arctic Ocean, partly because their newborn young remain, thinly coated and exposed to the wind for a month or more, on land. (This explanation, however, cannot be generally applied. *Histriophoca* has remained in the Pacific-Arctic while *Pagophilus, Halichoerus,* and *Cystophora* have remained in the Atlantic-Arctic, yet the newborn young in all of these genera are fat and woolly-coated, able to survive in icy waters.) Part of the answer to range restriction is perhaps that "border populations are tied by gene-flow to the integrated gene-complex of the main body of the species" and "adaptation by selection [at the border] is annually disrupted by the infiltration of alien genes and gene-combinations from the interior of the species range which prevents the selection of a stabilized gene-complex adapted to the conditions of the border region" (Mayr, 1954, pp. 163–66). The limiting factors of life near the border of a range may in fact be sharper and stronger than they seem. Little is known about the effects of ocean currents, upwellings, food abundance, ice thickness, ice movements, air temperatures, predation, disease, and the many other environmental influences in the life of the pinniped. Stefansson (1945, p. 101) learned from long experience that vast areas in the Arctic Ocean are "sea deserts" from the point of view of the seal hunter. These areas are often characterized by sluggish currents and thick, old ice.

Stabilization

The writer has suggested that Tertiary pinnipeds were closely attached to land and that they followed continental shores to remote parts of the earth, there to develop the stem-lines of the modern genera. Now he suggests that, as pinnipeds became more specialized and came to enjoy more fully the freedom of the sea, they fell increasingly under the influence of three factors which dampened the rate of evolution and brought stability. (The word "stability" is used in a relative sense. Evolution of pinniped subspecies and higher categories is of course still in progress today.) These factors were: increased *sociability,* increased *longevity,* and increased *mobility.*

1. *Increased sociability.* As compared with land carnivores, at least, the pinnipeds are now distinctly sociable. All seals are gregarious at breeding time. At one extreme (*e.g., Erignathus, Ommatophoca,* and *Hydrurga*) the breeding aggregations are small and diffuse, each numbering fewer than 100 animals. At the other extreme (*Callorhinus* and *Pagophilus*) the breeding aggregations are large and are concentrated on three or four islands or ice fields. Certain aggregations may include over one million animals.

What are the origins of sociability among seals? Where social groups have now become large and complex, as in *Callorhinus* and *Pagophilus,* common sense leads one to believe that the chances of survival of a "bunch-quitter" (a cattleman's term) are low. Especially would it seem to be advantageous for the young seal, facing a relatively long period of growth, to learn the location of feeding grounds and migratory pathways, and to acquaint itself with the general mores of its parent group. But how did social behavior start? Some possibilities may be reviewed (*a*) Vaz Ferreira has shown (*in lit.*) that South American sea lions tend to huddle together more compactly on the beach in cold weather than in warm. (*b*) James Fisher (1954, p. 81) expressed the opinion that "many sea-birds must be social feeders because a social flock is the best device for keeping in touch with a prey that lives in flocks or shoals." The phenomenon of seals concentrated at places of food abundance, as at the mouth of a river during a herring run, has been too often observed to need comment. (*c*) Seals move about with tides and currents, following prey, and accommodating to drifting pack ice. For weeks they may travel in fog and for months in polar darkness. A tendency to form aggregations, wherein an individual could more easily find a sexual partner and a mother could find her nursling, has obvious social value. (Sociability among seals does not seem to confer any advantage in *defense*. In fact, among flesh eaters in general—Carnivora, Pinnipedia, raptorial birds—there are few if any which band together to ward off other flesh eaters, though some of course may travel in packs for *offense*.)

2. *Increased longevity*. As stated on page 26, pinnipeds are long-lived. They are little affected by predation and disease. They tend to feed widely. They have the ability to fast, some for as long as four months. Together these attributes make for a population of many age classes—a population having genetic inertia.

3. *Increased mobility*. The term "migration" is conventionally applied to the seasonal movements of polar seals—seals that are absent *en masse* from the breeding site at one time of the year. However, the more is learned about seals, the more likely it seems that even those species dwelling in temperate zones move about rather freely, widely, and regularly. There is evidence, for example, that both California and Steller sea lions move north in winter (Fry, 1939; Bonnot, 1951, p. 374).

A sort of quantitative relationship between migratory habit and evolutionary stability has been studied in *Callorhinus*, a genus represented on breeding grounds in Alaska and Siberia at least 1,350 km. apart. Taylor *et al* (1955) have shown that thousands of Alaskan seals mingle with Asian seals in winter and spring off Japan, that subadult Alaskan seals bearing tags are occasionally recovered on Asian breeding grounds, and that specimens from Japanese waters are indistinguishable from those from Alaskan waters. One may conclude that intermingling during migration in winter results in some interbreeding later in summer (or, less likely, that in the period since *Callorhinus* split into eastern and western breeding populations there has not been time for anatomical distinction to assert itself).

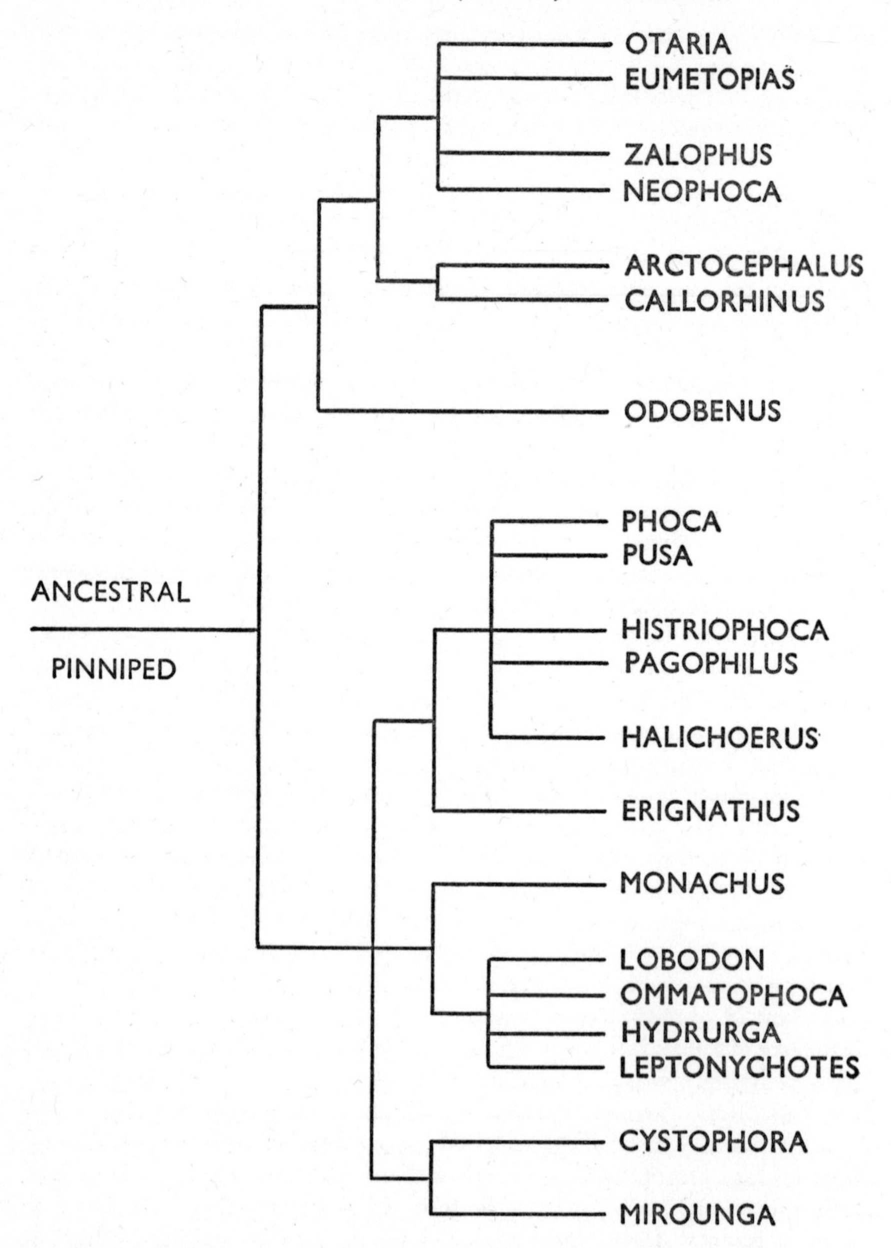

Fig. 1. Phylogenetic arrangement of the pinniped genera.

The stabilizing factors discussed in the preceding pages did not of course exert the same influence within all genera. They were important, for example, in the genus *Callorhinus* whose members are highly social (as many as 100,000 seen from one vantage point), and mobile (fanning out in winter over the North Pacific, with outriders to the Arctic Ocean, Mexico, and Japan). They were not important, on the other hand, in the genus *Phoca* whose members live in small groups of tens or hundreds without harem formation or other sharp social structure, and are resident more or less the year around. While *Callorhinus* is clearly monotypic, *Phoca* is clearly polytypic and in fact has been divided at one time or another into a score of species and subspecies.

Biotic factors (specializations) which have generally dampened the rate of evolution among pinnipeds have, in special cases, been nullified by strong geographical factors working in the opposite direction. Thus stabilizing developments within the genus *Zalophus* have not been able to prevent (in a manner of speaking) the evolution of distinct races on the Galapagos Islands and on the west coast of Mexico, separated by 23° of latitude. And the elephant seals *Mirounga* of the Northern Hemisphere have grown specifically distinct from their subantarctic relatives. Among the genera *Arctocephalus, Phoca,* and *Pusa* there has been—and still is—fairly rapid evolution, mainly because populations representing these genera tend to frequent isolated bays, gulfs, islands, inland seas, or lakes.

The effect of commercial sealing in the nineteenth century upon island populations was catastrophic. It is anyone's guess how soon certain islands of the Southern Hemisphere (*e.g.*, Isla San Ambrosio, Auckland Island, and the South Shetlands) will be reseeded with fur seals. Mayr (1954, p. 173) has stated that "the evolutionist takes, on the whole, a dim view of the future prospects of populations with depleted genetic variability. Such populations are not very plastic. If they live on an island . . . they will probably be successful as long as conditions remain stationary. However, such populations rarely have the capacity to adapt themselves to severe environmental shocks. The arrival of a new competitor or of a new enemy . . . is apt to lead to extinction. It is no coincidence that even though less than 20 percent of all species of birds are island birds, more than 90 percent of all bird species that have become extinct in historical times are island species. An island bird thus has at least fifty times as great a chance to become extinct as a mainland bird. Only part of this extinction can be attributed to the small size of the range . . ."

In the Pinnipedia the evolutionary pattern resulting, first, from adaptive radiation of generic stocks along separate, branching coastlines and, second, from final stability has been reflected in conventional taxonomic ar-

TABLE 3. RELATIVE NUMBERS OF GENERA, SPECIES, AND SUBSPECIES OF CARNIVORES AND PINNIPEDS IN SAMPLE FAUNAS*

	Order Carnivora:			Order Pinnipedia:		
Fauna	Genera	Species	Subspecies	Genera	Species	Subspecies
Palaearctic†						
Number	44	92	483	10	15	27
Ratio	*1.0*	2.1	11.0	*1.0*	1.5	2.7
Nearctic‡						
Number	27	73	439	11	14	21
Ratio	*1.0*	2.7	16.3	*1.0*	1.3	1.9
World§						
Number	...	...	...	18	28	0
Ratio	...	...	...	*1.0*	1.6	0
World¶						
Number	...	...	...	20	31	23
Ratio	...	...	...	*1.0*	1.6	1.1

* Purpose of the table is to compare within each fauna the number of genera with: (*a*) the number of species and (*b*) the number of subspecies.

† Ellerman and Morrison-Scott, 1951.

‡ Miller and Kellogg, 1955 (exclusive of the genus *Ursus* and 76 species assigned to it).

§ Allen, 1880.

¶ Present author.

rangements of the group. Historically the order Pinnipedia has been split into many genera but few species. In table 3 certain arrangements of Pinnipedia and Carnivora are compared. While the order Carnivora may not be an ideal group for reference it is clearly useful. A phylogenetic diagram of pinniped genera is offered in figure 1.

4

TAXONOMIC PROCEDURE

Historical Preface

Any attempt to classify the Pinnipedia is beset with special difficulties arising from the fact that few seal populations of the world have been adequately sampled. Many seals live on remote, oceanic islands or polar ice fields. All are large-bodied, greasy, bloody, and awkward to collect and to preserve for study. At the time of writing, the National Museum (for example) has no specimens representing the genus *Neophoca* and only one broken skull representing *Histriophoca.* The British Museum has no specimens representing *Histriophoca* and the species *Mirounga angustirostris* and *Monachus schauinslandi.* Certain seals of high commercial value had been nearly exterminated before zoologists were able to obtain specimens. On Islas Juan Fernández two or three million fur seals were killed; three skulls and one skin were saved to science; all but two skulls were subsequently lost by fire in World War II. The Juan Fernández seal is now extinct. Furthermore, age and sex variations reduce the effective size of any sample obtained (plate 1). In *Callorhinus* the adult male is about 4.5 times the size of the female, and 20 or more adult year classes of either sex may be present at one time and place. A grey seal *Halichoerus* dying at age 41 or 42 years in a zoo continued through life to demonstrate growth changes in profile of the head.

In reading the history of pinniped taxonomy one detects a change in sentiment from an earlier time when each new skin, skull, or sketch was considered grounds for a new species, to the present day when one specific name can represent a population distributed along a coast line of 10,000 to 20,000 km. (*e.g., Otaria byronia* and *Hydrurga leptonyx*). Allen (1880, p. 459) complained that in the Phocidae alone there were "probably nearly four hundred [distinct names] or an average of at least twenty names to each species, with a maximum for some of the species of at least thirty." And Turner (1888, p. 72) felt that "in no family of mammals, probably, have more diversities of opinion been expressed by zoologists, both with respect to the number of species in the family and their arrangement in genera and subfamilies, than in the Otariidae. These divergences are to be seen both in the descriptions of different authors and in those of the same author at different times."

Not until 1880 could Allen have written his remarkable monograph of North American pinnipeds, which traced the history of names, provided keys to the families and genera, described the North American species, and gave synopses of pinniped species of other parts of the world. His arrangement of the Pinnipedia into 18 genera and 28 species is still useful, though approximately half of his names have been modified and the subspecies concept has been introduced. While the order Pinnipedia has not been reviewed since 1880, instructive lists of pinnipeds have been contributed by von Boetticher (1934), Allen (1939, 1942), Bertram (1940), Cabrera and Yepes (1940), Doutt (1942), Bobrinskoi (1944), Simpson (1945), Carter *et al.* (1945), Anderson (1947), Dunbar (1949), Degerbøl (1950), Ellerman and Morrison-Scott (1951), Mohr (1952*b*), Ellerman *et al.* (1953), King (1954), Palmer (1954), Sivertsen (1954), Chapskiy (1955*a*), Frechkop (1955), Miller and Kellogg (1955), Rass *et al.* (1955), and Cowan and Guiguet (1956).

Pinniped nomenclature (as distinct from classification) has been reviewed since Allen by the following:

Trouessart (1897–1905). Genera, subgenera, species, and subspecies of mammals, living and extinct; authorities; synonyms; geographic ranges. Trouessart was especially interested in the Pinnipedia, of all mammals (1881, 1907, 1922).

Palmer (1904). Families, subfamilies, genera, and subgenera, living and extinct; authorities; type species and localities; etymology of names; other information. (Derivation of names has been more recently discussed by Brown, 1954, and Jaeger, 1955.)

Sherborn (1902–33). Genera and species, 1758–1850. Valuable bibliographies appeared in 1902 (Sectio Prima, pp. xi–lvi) and 1932 (Sectio Secundo, Epilogue, pp. cxxxiii–cxlvii).

Hay (1930). Check list of North American fossil seals, many of them recent species, with citations to world usage of names. For example, about 100 citations to use of the word "pinnipedia" and its equivalents are given.

Neave (1939–40, 1950). Genera and subgenera, 1758–1945.

Conisbee (1953). Continuation of Palmer's list, but confined to recent genera and subgenera, 1904–51.

Classification of the Pinnipedia is now, and will long be, equivocal. For example, the fur seal of Isla de Guadalupe was placed by one reviewer in 1954 in a certain genus and species; by another reviewer in 1954 in a quite different genus and different species. Nomenclature is also equivocal. For example, the distinct and long-known South American sea lion has been called *Otaria flavescens*, *O. byronia*, and *O. jubata* in three recent publications, respectively. The present writer concludes that, while classification of the Pinnipedia will always be to a certain extent subjective, it ought to be, first, based on a uniform scale of values; second, globally consistent;

and, third, conservative. To the last point the writer would add that stability is especially needed in pinniped taxonomy for the reason that the seals, sea lions, and walruses are of growing interest to man, and the sooner their scientific names become accepted in world language the sooner can nations agree upon international measures for conserving their populations.

Supergeneric Groups

A few additions or changes from the arrangement of Simpson (1945) have been made. Briefly, the Pinnipedia are regarded as an order rather than a suborder. The superfamilies Otarioidea and Phocoidea are added; also the subfamilies Otariinae and Arctocephalinae, the tribes Phocini and Erignathini, and the tribes Monachini and Lobodontini. The subfamily Monachinae (of Trouessart) has been restored. Supergeneric names are arranged and documented in the style of Simpson (his pp. 35–38). Here the valid name is the one first used with its present spelling and more or less its present meaning, but not necessarily its present rank.

Genera

Mayr *et al.* (1953, p. 48) have defined the genus as "a systematic category including one species or several species of presumably common phylogenetic origin, which is separated from other similar units by a decided gap." In the present review:

1. The genus is a group such that any individual member can be distinguished without question from a member of any other genus (provided that in certain cases specimens of the same age and sex are compared). At least one variate (measurement or ratio) pertaining to the skull does not overlap that of the nearest genus.

2. The genus is distinguished by a way of life of its members, especially habitat predilection, social structure, movements, food habits, and reproductive habits. Behavioral distinctions between pairs of allopatric genera, though, are not always conspicuous (*e.g., Eumetopias* and *Otaria*).

3. The principal breeding range of a genus is confined to one or two oceanic zones (see table 2, p. 39). The breeding ranges of two or three genera may overlap. That is, an aggregation of pinnipeds may include representatives of two or three genera side by side at breeding time (*e.g., Eumetopias, Zalophus,* and *Phoca; Lobodon, Hydrurga,* and *Ommatophoca*). The feeding range of a genus may extend into as many as four oceanic zones (*e.g., Arctocephalus*). The feeding ranges of as many as six genera may overlap, as previously mentioned on p. 39.

4. Members of one genus do not interbreed in the wild with members of another. (Nor do members of two species ordinarily interbreed if, by present definition, species are allopatric. See p. 49.) There are no records of intergeneric mating between pinnipeds in the wild. Even the closely

related and sympatric forms such as *Halichoerus, Phoca,* and *Pusa* remain reproductively apart. Intergeneric matings in captivity were mentioned on p. 25.

5. Finally as a clue to generic distinction one finds an indigenous name for the animal (*e.g.*, Kamchadal "lach-tak" and Canadian Eskimo "oogruk" for *Erignathus*; Canadian Eskimo "net-chek" for *Pusa*).

To illustrate application of the five foregoing criteria several examples may be given:

At one extreme, representing distinct genera, are *Zalophus* and *Eumetopias,* both members of the same subfamily; their ranges roughly end-to-end. Likewise distinct are the four genera of the subfamily Lobodontini. Four individuals representing *Lobodon, Ommatophoca, Hydrurga,* and *Leptonychotes,* each pursuing its own prey, might be seen at one time and place by an observer off the coast of Palmer Peninsula.

At the other extreme are *Neophoca* and *Zalophus,* similar in appearance but geographically separated by 9,000 km. As between *Neophoca* and *Zalophus,* Sivertsen (1954, p. 27, fig. 18) found only one diagnostic variate in the adult male skull. He had only 7 specimens of *Neophoca*; with a larger sample he would quite certainly have found overlapping. The present grounds for retaining *Neophoca* and *Zalophus* as separate genera are the sum of small differences in skull, the great distance between breeding ranges, and the intuitive feeling that when more is known about *Neophoca* its distinctions will be confirmed.

Other closely related genera are *Phoca, Pusa, Histriophoca,* and *Pagophilus.* All are smallish, spotted or banded seals of northern waters; placed by some writers in four subgenera of *Phoca.* But even the closest pair, *Phoca* and *Pusa,* would not for an instant be confused by an Eskimo. *Phoca* has a doglike face and a neutral odor, is associated with land, and feeds on a variety of fishes and shellfishes; *Pusa* has a catlike face and a rank odor (adult male), is associated with ice, and feeds extensively on macroplankton (except in landlocked lakes and seas).

As here classified, the order Pinnipedia includes 20 genera; 15 with a single species and 5 with more than one species (tables 1 and 2). And as previously noted, the percentage of monotypic genera is higher than among Carnivora (table 3). The splitting of genera in Pinnipedia has perhaps reached the point of maximum advantage. That is, one can now arrange the 20 genera so that gaps between related genera descend from maximum (*e.g.*, *Zalophus* and *Eumetopias*) to minimum (*e.g.*, *Zalophus* and *Neophoca*), while below the minimum one meets a decided gap before one encounters differences of specific importance (*e.g.*, species of *Monachus*; species of *Arctocephalus*). The writer agrees that the present arrangement stressing the importance of the genus may not be sympathetically received in all quarters. It is, however, the result of a personal conviction that twenty distinct kinds of seals inhabit the waters of the globe today, each long separated in geologic time from its nearest relative and each now circulating quite independently in its own biotic province.

In the systematic account starting on p. 52 the *type* shown for each genus is the species designated by the author of that genus or, where no type was designated, by a subsequent reviewer. Where the original name of the species is obsolete, its current synonym is given. Thus the type of *Zalophus* Gill is "*Otaria Gilliespii* Macbain" (= *Otaria californiana* Lesson).

Species and Subspecies

Among taxonomists the following opinions are widely held: first, that the species is the lowest category in which reproductive isolation is a criterion and, second, that subspecies are actually or potentially interbreeding groups, geographically contiguous or separate, whose members differ to the extent that the taxonomist can correctly assign about three out of four specimens to one group or another without knowledge of the source of the specimens. However, in recognition of the peculiar mode of evolution of the Pinnipedia, as compared with that of land mammals, and in recognition of man's still unsatisfactory knowledge of pinniped variation, the writer has applied to pinniped groups certain tests for species and subspecies which are felt to be useful and appropriate here, though perhaps not altogether to other mammalian groups. In estimating the relationship at trivial levels between two seal populations the following criteria have been kept in mind:

1. Species are clearly allopatric, are not part of a continuous distribution, and are fairly distinct in anatomy and behavior. Examples: ringed seal *Pusa hispida* and Baikal seal *P. sibirica*; South American fur seal *Arctocephalus australis* and Philippi fur seal *A. philippii*; southern elephant seal *Mirounga leonina* and northern elephant seal *M. angustirostris*. The thesis is presented that at any selected place in the marine environment there are too few ecological niches (providing food, protection, and breeding room) to permit the coexistence of two pinniped forms which are related as closely as species of the same genus. (On land, however, the situation may be different. For example, in Colorado the ranges of four *Mustela* species, namely, *erminea, frenata, nigripes*, and *vison*, are known to overlap.) Under terms of the present definition a situation reported by King (1954, p. 331) is regarded as accidental. "Both *A. australis* and *A. philippi* have been reliably recorded from Juan Fernandez, and *A. australis* from the Galapagos, so it is not unlikely, though not proven, that *A. philippi* also occurs on the Galapagos." The harbor seal, ringed seal, and harp seal, with overlapping ranges in the North Atlantic, often placed in the one genus *Phoca*, are regarded by the present writer (and in part by Chapskiy, 1955*a*) as members of separate genera. Where completely isolated forms have long been regarded as distinct species

and no fresh information about them has been produced, the writer has left the names in status quo.

2. Subspecies are allopatric, may or may not be part of a continuous distribution, and are slightly distinct in anatomy and behavior. Examples: *Phoca vitulina concolor* and *P. v. richardi; Zalophus californianus californianus* and *Z. c. wollebaeki; Arctocephalus australis australis* and *A. a. galapagoensis.* How to express taxonomically the differences which seem to be at, just above, or just below, the subspecific level is by general agreement a matter of opinion. The writer has elected to regard as conspecific those populations which are distributed along a more or less continuous line or circle, no matter how long. Take, for example, the broken circle with four radial extensions, totaling about 25,000 km., which represents the distribution of *Phoca vitulina.* A subspecific name is retained for each of the major segments of the line or circle where average differences have been reported. Thus *Phoca vitulina* is regarded as consisting of four subspecies, one along each radial extension (with its attached part of the polar distribution), plus a fifth (*P. v. mellonae*) in a landlocked lake.

3. Different populations of the same subspecies are allopatric, are probably in communication, and are not known to differ taxonomically in anatomy and behavior. Populations "in communication" are those in which a gene pattern is relayed from one end of the geographical range to the other in a matter of, say, a few centuries. Examples: *Lobodon carcinophagus* of the Weddell Sea and of the Ross Sea; *Otaria byronia* of the coasts of Uruguay and Peru; *Callorhinus ursinus* of the Pribilof Islands and the Commander Islands. *Callorhinus ursinus* is a climax example of a species whose representatives are widely separated on the breeding grounds but are in annual communication on the feeding grounds during migration. As previously stated, for this species the probability of interbreeding has been estimated through release and recovery of marked animals and through systematic comparison of specimens. Of other migratory species (*e.g., Pagophilus groenlandicus* and *Cystophora cristata*) similar studies are being carried on by Canadian, Norwegian, and Soviet investigators. It is extremely unlikely that any two species of the same genus will be found whose representatives consort on the feeding or molting range and later part for breeding purposes into groups of *morphologically* distinct animals. Some taxonomists, perhaps, would feel that *Callorhinus* and *Pagophilus* include sibling (physiological) species.

"The whole question of the appropriateness of the formal trinomial recognition of races [= subspecies], the ascription of a finite label to something which is not finite in nature, is in our view debatable," stated Ellerman *et al.* (1953, p. 2). Among the Pinnipedia, where a race is insular and finite (*e.g., Zalophus californianus wollebaeki*) the formal label is useful in identifying a transitional stage in the evolution of a species. But where a race is not finite (*e.g.,* many of the proposed subspecies along the distribution lines of *Phoca* and *Pusa*), ascription of a formal name may create the image of a discrete group which does not in fact exist. In an earlier paper (1955) the writer called attention to change in average size of fur seals during a 30-year period in response to a change in population pressure. Here the difference between first and third generations on the same island was greater than certain differences which, in other species, have led to the naming of races.

For each species and subspecies there is given: (1) A literature citation to the original use of the name with essentially its current spelling. The original description is to be found near the cited page or pages. (2) A citation to the name as first used and spelled in the present binomial or trinomial combination. (3) Where particularly useful, citations to other names applied to the species or subspecies; or a citation to a published history of the name. For additional synonyms the reader is referred to Trouessart's *Catalogus* (1897–1905), the most recent work treating systematically the pinnipeds of the world. (4) Under "type," a description of the material which led the original author to conclude that he was recognizing a new and hitherto undescribed kind of pinniped. For species named before 1850 (or thereabouts) the material was often little more than a seafarer's account; for later species, a type specimen or series of specimens—actual skins and bones. Types are defined as follows (*cf.* Frizzell, 1933; Simpson, 1945; Follett, 1955; Schenk and McMasters, 1956; where "species" is given, read "and subspecies"): The *type* of a species is the single specimen upon which the species was based, and which was either designated as the "type" or the "holotype" by the author or was undesignated but was clearly the sole specimen before him. The *syntype* of a species is one of several specimens of equal rank used by the original author as the basis of the species when no type was designated. A *paratype* is a (subordinate) specimen, other than the designated type, upon which the original description was based. (Opinion today is against the designation of syntypes and paratypes in systematic mammalogy.) The *lectotype* is a single specimen selected by the author or other reviewer, after the original publication, from a series of syntypes, to be the name bearer of the species. (5) Type locality and geographic range, north to south and east to west. The locality name is spelled in accordance with recommendation of the U.S. Department of the Interior, Board on Geographic Names. Where two locality names are given one after the other and the second is in parentheses, the second is a subdivision of the first. Example: Islas de Castillo Grande (Isla del Marco). However, where the second is preceded by an = sign it is a synonym of the first. Where the Board gives two alternatives, the conventional spelling (*e.g.*, Cape Horn) is used in preference to the foreign (*e.g.*, Cabo de Hornos). (6) Vernacular names, for species and higher groups, offered with the realization that these will not be familiar and acceptable in all parts of the English-speaking world.

5

SYSTEMATIC ACCOUNT*

Order PINNIPEDIA Illiger, 1811

Pinnipedia Illiger, 1811, p. 138. (seals, sea lions, and walruses)

REMARKS. Storr (1780, p. 41) had earlier erected the phalanx "Mammalia Pinnipeda" to include all pinnipeds and sirenians known to him. Illiger, however, was first to recognize the pinnipeds as distinct. He introduced "Pinnipedia (Ruderfüsser)" as name both for an order and a family.

The pinnipeds are an ancient group–isolated, aberrant, unique. At a glance one can distinguish a seal or walrus from any other kind of mammal. Zoologists generally agree that pinnipeds, in spite of resemblances to certain members of the order Carnivora, deserve recognition as a parallel and independent order.

Superfamily OTARIOIDEA Smirnov, 1908

"Gressigrada s. Otarioidea" Smirnov, 1908, pp. 1, 14. (gressigrade seals)
Otarioidea Gregory and Hellman, 1939, p. 313.

REMARKS. The two primary groups of pinnipeds: "walkers" and "wrigglers" were named by Allen (1880, pp. 3–4) Gressigrada and Reptigrada. Smirnov substituted Otarioidea and Phocoidea, based on the names of included genera.

Family OTARIIDAE Gill, 1866

Otariidae Gill, 1866*a*, pp. 7, 10. (otariids, eared seals)
Otariadae Brookes, 1828, p. 37.

REMARKS. "Péron, in 1816, first divided the Seals into two genera, separating the Eared Seals from the earless ones under the name Otaria. Later, Brookes . . . raised the group of Eared Seals to the rank of a family, under the name of Otariadae. This classification was not, however, generally adopted till 1866, when it was revived by Gill, and immediately adopted by Gray, and it has been accepted by most subsequent writers" (Allen, 1880, p. 188).

* For plan of the account see preceding chapter; for terminology of measurements see p. 135; for phylogenetic diagram see figure 1 on p. 42.

Subfamily OTARIINAE von Boetticher, 1934

Otariinae von Boetticher, 1934, p. 359. (sea lions)

Otariina Gray, 1825, p. 340, as a "tribe" of the "family Phocidae"; the tribe including *Otaria* and *Platyrhynchus* (= *Otaria*).

REMARKS. Allen (1870, p. 23) split the eared seals into two groups, stating that "the name Trichophocinae is proposed for the hair seals [sea lions] and Oulophocinae for the fur seals." Von Boetticher rejected Allen's names and identified the sea lion and the fur seal groups with the names of included genera. He also introduced a third category: the "Mittelrobben Phocarctinae" to contain only *Phocarctos hookeri* (= *Neophoca hookeri*), presumed to have neither true fur nor true hair. Simpson (1945, p. 233) stated that "in the absence of a consistent arrangement of all the otariid genera it is inconvenient to recognize any isolated subfamilies." However, the present writer feels that the living members, at least, fall naturally into two fairly distinct groups, here treated as subfamilies.

No otariid has pelage intermediate between that of the sea lion (with overhair only) and that of the fur seal (with a distinct layer of underhair as well as overhair). Skins of *Neophoca cinerea* and "*Phocarctos hookeri*" (= *Neophoca hookeri*) in the British Museum (Natural History) are clearly those of sea lions, not fur seals. The sea lion type of pelage may represent a mutation such as the "samson" sport of the red fox of Finland. According to Voipio (1956, p. 17), the samson strain of *Vulpes fulva* has only one layer of hair instead of the normal two. The strain probably entered the stock at a moment when the fox population was low or near extinction. Samson foxes are intolerant of cold and are unknown in northern Finland. Similarly, the breeding centers of sea lions, both in the Northern and Southern Hemispheres, are slightly nearer the equator than are those of fur seals.

Genus **OTARIA** Péron, 1816

Otaria Péron, 1816, vol. 2, p. 37, footnote (part), pp. 40–52 (part).

TYPE. *Phoca byronia* Blainville.

REMARKS. In his account (published posthumously) Péron proposed a "famille nouvelle" *Otaria* to include the eared seals. Allen (1905, p. 101) decided that the type by restriction is *Phoca byronia*. Frechkop (1955, p. 320) considered the type to be "*O. jubata* (Forster, 1775)" although in footnote he called attention to Sivertsen's (1953*b*) use of *O. byronia*. For bibliographic purposes the present writer lists the type of *Otaria* as *Phoca byronia* Blainville and the type of *Eumetopias* as *Phoca jubata* Schreber. The troubled history of the word "*jubata*" will be discussed under *Eumetopias*.

Otaria byronia (Blainville) 1820. (South American sea lion)

"*P*[*hoca*] *Byronia ?*" Blainville, 1820, p. 300, fig. 3. South America.

"? *Otaria byronia,*" Peters 1866*a*, p. 269.

Otaria byronia, Trouessart, 1897, p. 369 (*O. flavescens* Shaw, 1800, p. 260, placed in synonymy)

TYPE. *Holotype*: An adult male skull, lacking lower jaw and most of the teeth, brought to England by Commodore John Byron in 1769 and placed in the British Museum; purchased in 1809 by the Museum of the Royal College of Surgeons of England; listed in Flower's catalogue (1884, p. 189) under *O. jubata* as Osteological Collection no. 3966. It was destroyed during World War II. The skull was certainly that of *O. byronia.* Labeled from Tinian Island, Caroline Islands (where there are no pinnipeds), it probably came from the Strait of Magellan or Islas Juan Fernández, both visited by Byron (Allen, 1905, p. 112; Hamilton, 1934, p. 273).

RANGE (fig. 2). *General*: Brazil to Strait of Magellan and Peru; Falkland Islands. *Brazil*: Vaz Ferreira saw 200-300 sea lions on Isla de Torres, between Estado de Santa Catarina and Estado de Rio Grande do Sul in April 1953 (*in lit.*). Sporadically to coast of Rio de Janeiro (about 23° S) (Cunha Veira, 1955, p. 456). *Uruguay*: Islas de la Coronilla, Islas de Castillo Grande, Islas de Torres, Isla de Lobos, Rocas de las Pipas; sporadically as far as 300 km. up the fresh-water Río de la Plata; local populations fluctuating greatly with the seasons (Vaz Ferreira, 1956*a* and *in lit.*). That author estimated 44,000 sea lions on Isla de Lobos (1950, p. 155). *Argentina*: About 1,500 in a colony near Santa Cruz, 50° 01′ S, 68° 31′ W (Brandenburg, 1938). Santiago Carrara (1952, appendix pp. 3-5) listed 66 "loberias" along the coast of Argentina with estimated total population of 143,905 animals. About 4,000 skins were taken annually between 1949 and 1951. In 1954 he raised the estimate to 168,270. *Pacific coast of South America*: Verrill (1870, p. 431) stated that "in the Museum of Yale College there is a large adult skull of [*Otaria*] collected by Prof. F. H. Bradley much farther north, at Zorritos, Peru (about lat. 5° S) [actually 4° S] where the marine fauna is eminently tropical and nearly identical with that of Panama." This seems to represent a northern limital record. Milton J. Lobell, for several years a Fish and Wildlife Service fisheries specialist on the west coast, wrote recently (*in lit.*) that sea lions are abundant along Peru, especially the southern part. He has seen them as far north as Paita (5° S). Small industries are based on the killing of sea lions for hides, oil, and meat for fertilizers or feed. Bini (1951) stated that sea lions occur to Paita or a little farther north. Kellogg (1942, p. 455) estimated that 75,000 skins were being taken annually from rookeries along the Peruvian coast. Peters' type of "*Otaria*

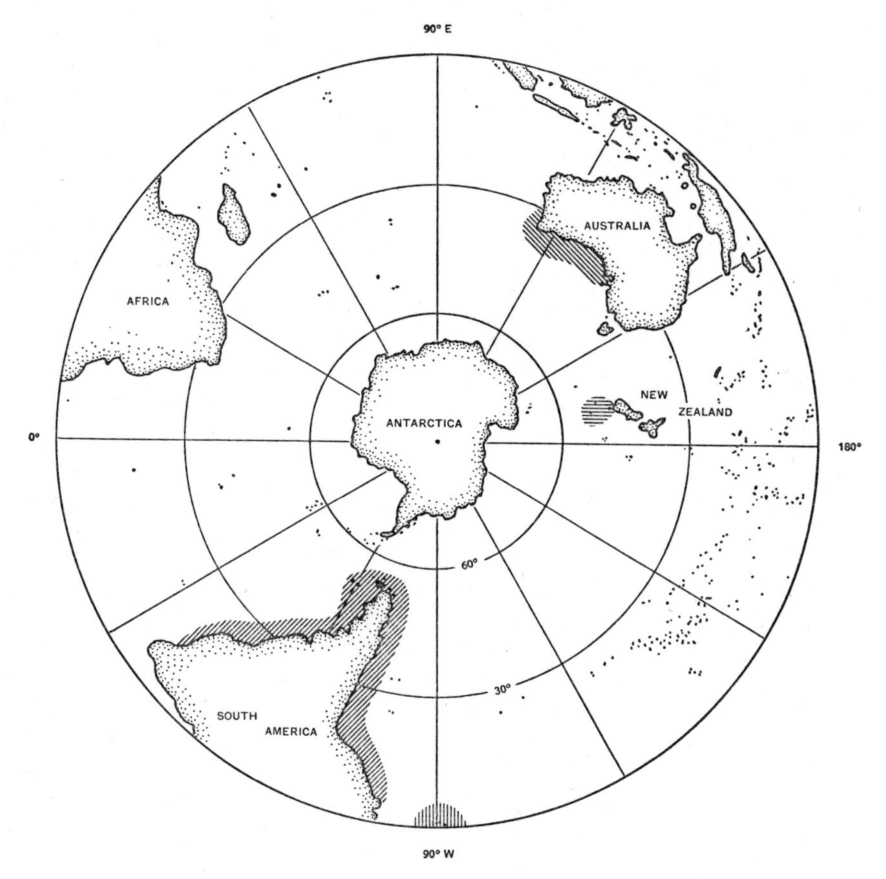

Fig. 2. Ranges of sea lions in the Southern Hemisphere. (///) South American sea lion *Otaria byronia*. (\\\) Australian sea lion *Neophoca cinerea*. (≡) New Zealand sea lion *Neophoca hookeri*. (|||) California sea lion (part) *Zalophus californianus wollebaeki*. (On this and the following maps an attempt has been made to show main breeding ranges as they exist today. Limital records and notes on migration are given in the text.)

Godeffroyi" (1866*a*, p. 266) was collected in the Islas de Chincha. The skeleton and mounted skin were lost by fire in 1943 (Mohr, 1952*a*, p. 110). Chapin (1936) found sea lions here and on Hormigas de Afuera, near Callao. Murphy (1920, p. 73), traveling southward along the coast of Peru, found sea lions and a sudden drop in seawater temperature at Pariña, 4° 45′ S. Osgood (1943, p. 100) gave the range of *O. byronia* "from the Galapagos Islands southward along the entire Chilean coast to Tierra del Fuego . . . Recent reports from Juan Fernandez [= Isla Más a Tierra] and other islands off the coast of Chile are lacking." Sivertsen (1954, pp. 39–40) has shown, however, that *Otaria* is not recorded from the Galapagos. Many years earlier Allen (1880, p. 208) had included the Galapagos in the range of "*Otaria jubata*" on the basis of specimens from the Hassler Expedition. With regard to these specimens Barbara Lawrence Schevill wrote (*in lit.*) that "our two so-called '*Otaria byronia*' from the Galapagos are incomplete skeletons lacking skulls, and one of them is young . . . I am sure they are not *Otaria.* My feeling is that they are *Arctocephalus galapagoensis.*" *Falkland Islands*: Hamilton (1939*b*, pp. 160-61) estimated 380,000 sea lions on the Falkland Islands in 1937. Matthews (1952, p. 88) stated that sea lions were increasing rapidly and were being killed on fur-seal rookeries to remove competition. Laws (1953*a*, p. 96) estimated 400,000 sea lions here. The species has not been reported south of the Falklands.

REMARKS. Use of the name "*O. flavescens*" was defended by Cabrera (1940, pp. 17–22; and *in lit.*). From his experience with South American pinnipeds he concluded (rightly, no doubt) that the "Yellow Seal –*Phoca Flavescens*" of Shaw can have been only a southern sea lion pup after its first molt. Osgood (1943, p. 101) agreed. The present writer, however, prefers the name *O. byronia.* As against *O. flavescens* it is based on a tangible specimen. A detailed history of names was given by Allen (1905, pp. 105-8) and by Osgood (1943, pp. 99-100).

Genus **EUMETOPIAS** Gill, 1866

Eumetopias Gill, 1866*a*, pp. 7, 11.

TYPE. *Phoca jubata* Schreber.

REMARKS. The type designated by Gill was "*Otaria californiana* Lesson., = *Arctocephalus monteriensis* Gray." Since the former was shown by Allen (1880, pp. 292-93) to be *Zalophus,* only the latter is available as type. See discussion of "*jubata*" under species. The type of *A. monteriensis* is a skull only, adult male, British Museum (Natural History) no. 1859.11.5.1, original no. 1320*a;* collected at Monterey by J. H. Gurney and figured by Gray (1859*b*, pl. 72).

Gill regarded the word "*Eumetopias*" as masculine and linked it with "*californianus*" (1866*a*, p. 13). A few other writers (*e.g.*, von Boetticher, 1934; Soviet taxonomists; and King, 1954) have followed his example by linking it with the masculine form "*jubatus.*" But "*Eumetopias*" resembles certain Greek words which are feminine, and though itself not a classical word, it should, by analogy, be regarded as feminine. This is the opinion of J. A. Crook (Fellow and Tutor, classical languages, St. John's College, Cambridge, *in lit.*) and most recent reviewers.

Eumetopias jubata (Schreber) 1776. (Steller sea lion)

Phoca jubata Schreber, 1776, Theil 3, p. 300 and pl. 83*b*. Bering Island, Commander Islands. Description on p. 300 of the sea lion of Steller; name *Phoca jubata* as caption of colored plate 83*b*, separately bound. References to Pernetty, and the illustration itself, do not apply. Schreber's earlier volumes were undated; Sherborn (1902, p. 503) gave 1776 as the date both for p. 300 and pl. 83*b*. Schreber's *Phoca jubata* was thus composite, based mainly on the non-Linnaean "Leo marinus" of Steller (1751, p. 360) and partly on the non-Linnaean "Lion marin" of Pernetty (1770, Tome 2, pp. 37–51 and pl. 10 at end). Forty years later Péron (1816, pp. 35, 53) restricted the name *jubata* to the sea lion of Steller, and here it seems wise to leave it.

Otaria stellerii Lesson, 1828, p. 420. Bering Island.

Arctocephalus monteriensis Gray, 1859, p. 358. A fetal skull from Monterey, California (= *Eumetopias jubata*). Figured by Gray (1872, figs. 4–5) with caption "*Eumetopias stelleri.*"

Eumetopias jubata, Allen, 1902*a,* p. 113.

TYPE. None. Species based on Steller's observations, starting 20 June 1742, of a sea lion rookery on Bering Island, Commander Islands. His drawing of a sea lion was lost although his notes were published in 1751 in *De Bestiis marinis* (Stejneger, 1936). The range given by Steller was from 56° N, southward "along the American shores" and Kamchatka to "Matmej Island" (= ? Matsua-tō or Ostrov Matua, 48° 05′ N, 153° 13′ E, in Kuril Islands).

RANGE (fig. 3). *General*: Breeding from northeastern Bering Sea (Pribilof Islands), Aleutian Islands, and west coast of North America, southward to southern California, westward to Commander Islands, Kamchatka, and Japan. *American waters*: Thousands breed on the Pribilof Islands, Bogoslof Island, Aleutian Islands, and the Alaskan-Canadian coastline. Imler and Sarber (1947, p. 7) found 4,000 in the Barren Islands, near Kodiak Island. Kenyon and Scheffer (1955, p. 11) estimated the American populations at: Alaska 40,000; Canada 10,000; Washington 500; Oregon 1,000; and California 3,000; total 54,500. *Asian waters*: Barabash-Nikiforov (1938, p. 427) stated that in the western Pacific "near the Commander Islands sea lions, almost exclusively males, occur only in the winter period from October until April. From the beginning of April they gradually disappear and probably migrate to the east coast of Kamchatka. One cannot find their young on the Commander Islands . . . In summer only solitary young male sea lions live near the islands. On Copper Islands these males often associate with the fur seals." Naumov (1933, p. 21) stated that sea lions do not occur in the southern part of Sea of Okhotsk or in

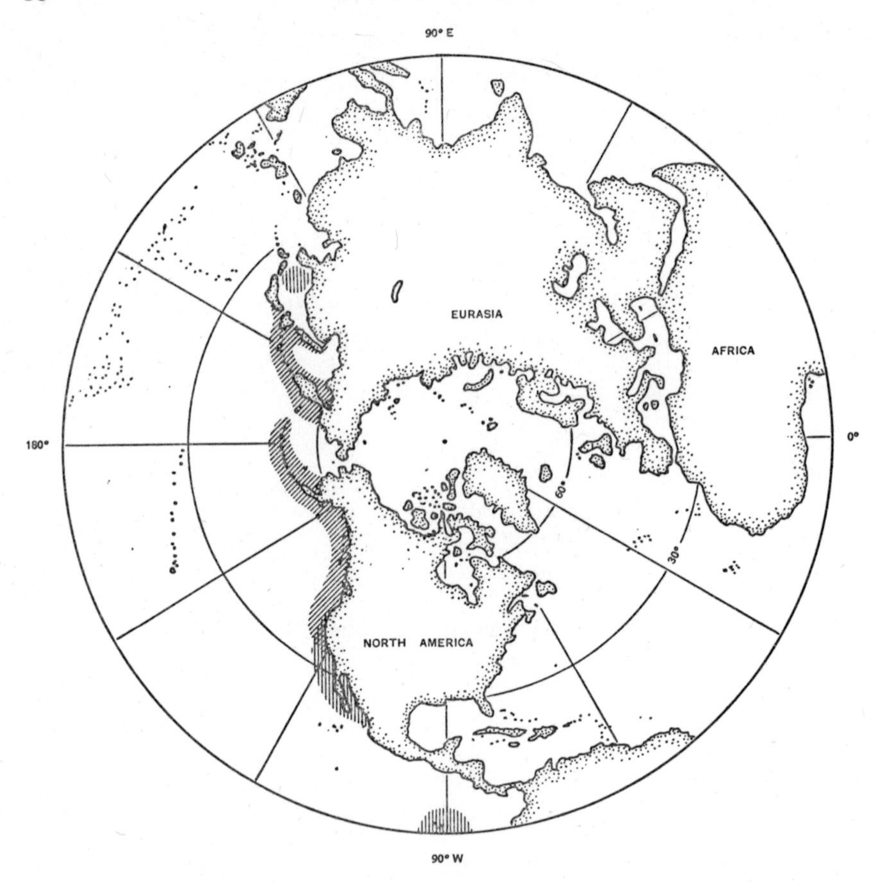

Fig. 3. Ranges of sea lions in the Northern Hemisphere. (///) Steller sea lion *Eumetopias jubata.* (|||) California sea lion: *Zalophus californianus californianus,* California to Mexico; *Zalophus californianus japonicus,* Sea of Japan; *Zalophus californianus wollebaeki,* Galapagos Islands.

the northern part of Tatar Strait but are widely scattered in the Sea of Japan. The largest Siberian rookeries are in Kronotskiy Zaliv (54° 12′ N, 160° 30′ E), on Ostrova Yamskiye (59° 12′ N, 155° 18′ E), and on Ostrov Svyatov Iony (56° 24′ N, 143° 27′ E). About 1,500 sea lions are killed yearly in the U.S.S.R. Nikulin (1937*a,* p. 48) stated that there are large rookeries in the Sea of Okhotsk: 5,000 to 6,000 on Ostrov Svyatov Iony; others on Ostrov Yamskiye and on Mus Shipunskiy (53° 04′ N, 160° 01′ E) on the east coast of Kamchatka. Okada (1938, p. 19) reported sea lions extending from Sakhalin and the Kuril Islands to Hokkaido, Honshu, and Korea. His authority for including Korea is perhaps the statement of Fauvel (1892, p. 455) that "quatre variétés de phocidés" are known from the northern China coast. Rass *et al.* (1955, p. 101) published a map of the distribution of *E. jubata* in western Bering Sea and Sea of Okhotsk. *Limital range*: Yukon Territory, Herschel Island (69° 35′ N, 139° 05′ W) (McEwen, 1954), Bering Strait (Pohle, 1932, p. 78), southern California, breeding from Santa Rosa Island (33° 57′ N) northward (Bonnot, 1951, p. 374; Collyer and Baxter, 1951), Korea. The widest gap in its range is apparently the 650 km. distance between the Aleutian Islands and Commander Islands. Steller sea lions move along the coast in the intervals between breeding, though little is known about their movements. Bonnot (1951, p. 374) concluded that sea lions move northward from California in winter since "very few Steller bulls can be found" at this season. Like other sea lions, *E. jubata* is occasionally seen in rivers. One was captured in a pasture near Oregon City, Oregon, over 150 km. from the ocean (Weed, 1936).

REMARKS. The history of the word "*jubata*" is apparently as follows: Between 30 December 1774 and 2 January 1775, the sloop *Resolution* under Capt. James Cook lay off Isla de los Estados near Cape Horn. Aboard were J. R. Forster, naturalist, and his son George. During this period a number of sea lions (*Otaria byronia*) were killed. George Forster (1777, p. 515) remarked that "the late professor Steller found these animals at Bering's Island near Kamchatka . . . and his descriptions, the first and best ever drawn up, correspond exactly with our observations." Many years later, in a posthumous work by the elder Forster (1844) the word "*jubata*" appeared in a table on p. 66. It was used on p. 317 in the combination "*Phoca iubata* F." formally to describe the sea lion of Isla de los Estados, seen on 2 January 1775, and thought by Forster to be identical with the sea lion of Steller. Thus Forster did not use the word "*jubata*" before 1844, and the combination "*jubata* Forster, 1775," still occasionally seen, is incorrect.

In view of the contribution of Steller to knowledge of the northern sea lion it is unfortunate that the name "*stelleri*" was knocked out by the law of priority. However "*jubata*" is well established (*e.g.*, Bobrinskoi, 1944; Ellerman and Morrison-Scott, 1951; and Miller and Kellogg, 1955) and will undoubtedly continue to be used.

Genus **ZALOPHUS** Gill, 1866

Zalophus Gill, 1866*a*, pp. 7, 11.

TYPE. "*Otaria Gilliespii* Macbain" (= *Otaria californiana Lesson*).

REMARKS. According to J. A. Crook (*in lit.*) and the international rules (Follett, 1955, p. 10, par. 65A) the word "*Zalophus*" should be regarded as masculine.

Mohr (1952*b*, p. 161) described successful matings in captivity between "*Otaria pusilla*" male and "[*O.*] *californiana*" female (= *Arctocephalus pusillus* × *Zalophus californianus*). Young were born from 1911 onward, at least one of which was successfully reared. These matings represent interfertility between members of two closely related subfamilies.

Zalophus californianus (Lesson) 1828. (California sea lion)

Otaria californiana Lesson, 1828, vol. 13, p. 420. San Francisco Bay, California.

Zalophus californianus, Allen, 1880, pp. 205, 276.

Otaria gillespii [*lapsus* for *gilliespii*] M'Bain, 1858, p. 422. Probably the northern end of Gulf of California, Mexico; type skull figured by Turner (1912, p. 171); specimen now University of Edinburgh Anatomical Museum no.P.EU.ca 1.

TYPE. None. Species based on the statement of Choris (1822, p. 7) that "les rochers dans le voisinage de la baie San-Francisco sont ordinairement couverts de lion marins. Pl.XI." The plate shows a "Jeune lion marin de la Californie"; a small, grayish-brown otariid.

RANGE (figs. 2 and 3). Three isolated populations, no longer in communication and now regarded as subspecies, inhabit the west coast of North America, the Galapagos Islands, and the southern Sea of Japan.

Zalophus californianus californianus (Lesson) 1828. (See origin of name under species.)

RANGE. *General*: Southern British Columbia (49° N) to Mexico (21° 30′ N). *British Columbia*: "California sea lions occur almost every winter in the Barkley Sound area on southern Vancouver Island. Specimen records are from Ucluelet [49° N]" (Cowan and Guiguet, 1956, p. 350). *Washington and Oregon*: "During the winter months . . . regularly found northward to central Oregon and perhaps to the mouth of the Columbia River . . . Two adult males were seen on Jagged Island, off the northern Washington coast on 11 May 1954" (Kenyon and Scheffer, 1955, pp. 14–15). *California*: Breeding from Piedras Blancas (35° 40′ N) southward to the Mexican border; 9,000 sea lions (including Steller sea lions) on California coast (Bonnot, 1951). *Mexico*: Islas los Coronados, Isla de Guadalupe, Islas San Benito (9,500 estimated in February 1950; Hubbs, 1956*a*, p. 19), and many rocks off the coast of Baja California. There are numerous records of sea lions in the Gulf of California, and the type specimen of *Otaria gilliespii* was found by a seaman "at the mouth of the Red River, which we infer to be the Rio Colorado at the head of the gulf" (M'Bain, 1858, p. 428). When Dampier (1937, pp. 182–83) sailed up the Mexican coast he first encountered seals on 20 January 1686 on the east side of "Islands Chametly . . . six small islands, in lat. 23d. 11 m. a little to the South of the Tropick of *Cancer*, and about 3 leagues from the Main." (See also Burney, 1816, vol. 4, pp. 223–24, 228.) These animals were quite certainly California sea lions. The southern limit of the range is (or was) Islas Tres Marías (21° 30′ N). Nelson (1899) saw a single sea lion here and believed that a few were living in each of the three islands: María Cleofa, María Madre, and María Magdalena. Hanna (1926) did not find seals of any kind among the Islas Revillagigedo, 18°–19° N, during a careful survey in April and May 1925.

REMARKS. *Zalophus c. californianus* is often displayed as the "trained seal" of circus or zoo. Of 12 sea lion births in London Gardens, England, all occurred between mid-May and mid-June (Zuckerman, 1953, p. 853). These dates correspond to the pupping season of the wild parent stock, though the stock must have come from a latitude 20° south of London.

Zalophus californianus japonicus (Peters) 1866. New combination.

Otaria japonica Peters, 1866*b*, p. 669. Sea of Japan.

Zalophus lobatus, Jentink, 1892, p. 115. The author listed "a. Mâle adulte monté, un des types du *Otaria stelleri* Temminck. Japon. Des collections de M. Bürger. *Otaria japonica* Schlegel. b.c. Jeunes individus montés. Japon. De M. Bürger."

Zalophus japonicus, Carter *et al.,* 1945, p. 100. On the basis of Temminck's color plate (1847, pl. 21) and description of the animal as "straw colored with a darker throat and chest in the female" the Japanese form was regarded as distinct.

TYPES. History of the syntypes is uncertain; see *Remarks.*

RANGE. Apparently confined to the Sea of Japan; about 8,500 km. from the nearest breeding colonies in California. Uncommon? Karl W. Kenyon and Ford Wilke (Fish and Wildlife Service) visited two zoologists in Tokyo in 1952: Nagamichi Kuroda and Tadamichi Koga (the latter director of Ueno Zoological Gardens). The following information was obtained:

(Kuroda.) A sea lion, not *Eumetopias,* lives on Take-shima (= Liancourt Rocks, 37° 15′ N, 131° 52′ E) in Shimane-ken, Sea of Japan. This is a rocky islet. According to a Japanese newspaper 50 to 60 sea lions were seen here in November 1951. The animals are said to be increasing and spreading along the coast of Shimane-ken. The species formerly lived on Kyushu, Shikoku, Seto-naikai (= Inland Sea), Sagami-wan (35° 15′ N, 139° 25′ E), and Kōzu-shima (34° 13′ N, 139° 10′ E) in Izu-shichitō. No specimens are known in Japanese museums.

(Koga.) Kynkichi Kishida, who saw the animal twice on Tori-shima (where? there are four islands of this name), wrote a paper in Japanese about 1930 using the name *Zalophus lobatus.* (The present writer has not seen Kishida's paper. Kishida's name is listed twice in the bibliography of

Kuroda, 1938, p. ii.) The distributional range was given as Tsugaru-kaikyō, between Honshu and Hokkaido, southward to Australia.

Published accounts by Kuroda (1938, p. 23) and Okada (1938, p. 19) do not add materially to the foregoing information. Dr. Fukuzo Nagasaki told the writer in 1958 that there are "more than 200" California sea lions in the Sea of Japan. Some breed on islands claimed both by Korea and Japan.

REMARKS. Peters (1886*b*, p. 669) found in the Leiden Museum the skull and skin of "*Otaria japonica* Schlegel Mspt. (*O. Stelleri* Schleg. non Lesson) aus Japan." He identified these with *Otaria gillespii* M'Bain (= *Zalophus californianus* [Lesson]). The name "*japonica*" was apparently written by Schlegel on the labels of specimens assembled by him (in 1847?). The specimens were later described by Temminck (1847, pp. 10–12, pls. 21–23) under "*Otaria Stelleri.*" Most of the specimens are actually *Zalophus*; one or two(?) are *Eumetopias,* according to Sivertsen (1954, p. 35, and *in lit.*). A. M. Husson, Curator of Mammals, Rijksmuseum van Natuurlijke Historie, Leiden, wrote (*in lit.*) that the Museum holds a mounted specimen without skull, labeled "*Otaria japonica* nov. spec.? *Otaria mollissina* Lesson?" without the name of Schlegel but noted as collected in Japan by D. W. Bürger. Trouessart did not give "*japonicus*" as the name of any pinniped.

With some doubt as to the validity of "*japonicus*" the combination *Zalophus californianus japonicus* (Peters) 1866 is presently used to identify the California sea lion of Japanese waters. Only 8 specimens are known to the writer. Sivertsen examined 7 of these: 1 in the British Museum (Natural History) and 6 in the Rijksmuseum van Natuurlijke Historie. The British Museum skull (1873.3.12.1) is large; certainly that of an old male: CBL 323, BLH 292, MW 173, ZW 187, IOW 37, LUTR 111, height of sagittal crest 38. Sivertsen found this skull much larger than any from California (max. CBL 309) and larger than two adult males from Japan in the Temminck collection (CBL 307 and 207). An eighth specimen is also in the British Museum (1884.4.15.1), from Yokohama. It is a skin and skeleton with fragmentary skull, apparently an adult but not full-grown female. The dried skin is about 164 cm. (45.7 in.) in length, grayish brown with darker gray along back.

Kishida (according to Koga, *supra*) gave the lengths of adult male and female as 2.5 m. and 1.8 m. A 4-month-old pup was said to weigh 9 kg. and to measure 65 cm. in length.

Zalophus californianus wollebaeki Sivertsen, 1953. New combination.

Zalophus wollebaeki Sivertsen, 1953*a*, p. 2. Isla Santa María, Galapagos Islands.

TYPES. *Holotype*: Skull only, adult male, Zoologisk Museum, Oslo, no. 7931; collected 12 December 1925 on Isla Santa María (= Floreana Island), Galapagos Islands, by Alf Wollebaek. *Paratype*: Stuffed skin and skull of young female, no. 7932, collected 25 August 1925 on Isla San Cristóbal (= Chatham Island) by Wollebaek (Sivertsen, *in lit.*).

RANGE. Baur (1897, p. 956) gave names of nine Galapagos Islands on which sea lions were seen "in considerable numbers, especially on Hood and Gardner, Barrington and Duncan" (= Isla Española, Bahía de Gardner on Isla Española, Isla Santa Fe, and Isla Duncan). Heller (1904, p. 244) found sea lions "on all the islands of the Galapagos Archipelago . . . This is the most abundant seal in the Archipelago and breeds whereever found. The breeding season does not appear to be confined to any definite time of the year, as pups of all ages were found at all the rookeries during our stay of six months [read five months? January to May, 1899]." Specifically, sea lions were seen on Narborough Island (= Isla Fernandina), Albemarle (= Isla Isabela), Seymour (= Isla Seymour), Hood (= Isla Española), and Barrington (= Isla Santa Fe). The largest series of rookeries extended for about 15 miles on mangrove-fringed lagoons in quiet waters along the east coast of Isla Fernandina. As mentioned above, the types of Z. c. *wollebaeki* were collected on Isla Santa María and Isla San Cristóbal. Heller and Robert E. Snodgrass, who accompanied the sealing schooner *Julia E. Whalen* to the Galapagos Islands, found sea lions so abundant and tame that the five sailors of the crew took 2,100 skins in about four months.

REMARKS. Sivertsen (1953*a*, 1954) has described in detail the differences between the California sea lion of the North Pacific and the one of the Galapagos Islands. The differences seem to the present writer of subspecific rather than specific importance. In general, Z. c. *wollebaeki* is smaller in both sexes, the skull is more slender (except that the brain case at the jugals is wider), the upper postcanines number 5 - 5 in 79 percent of Galapagos skulls but 6 - 6 in 75 percent of North Pacific skulls. The greatest difference, in a comparison of adult male skulls, 20 from the north and 23 from the south, is the index:

$$\frac{\text{width at preorbital processes}}{\text{width of brain case}} \times 100$$

The index for northern skulls ranges from 93 to 123; for southern skulls from 73 to 88.

Heller (1904, p. 245) described the adult female, wet; upper parts dusky gray; throat and sides of neck and body light yellowish brown; chest and belly brown; nose black; vibrissae pale. The largest of 3 adult females measured (mm.): length 1,437, tail vertebrae 81, ear 31, length of vibrissae 156.

Genus **NEOPHOCA** Gray, 1866

Neophoca Gray, 1866*a*, p. 231. (Tasman sea lions)

TYPE. *Arctocephalus lobatus* Gray (= *Otaria cinerea* Péron).

REMARKS. The sea lions (Otariinae) presently listed under *N. cinerea* and *N. hookeri* are often confused with fur seals (Arctocephalinae). As recently as 1934, von Boetticher proposed a subfamily "Phocarctinae" to contain the sea lion of Auckland Islands and Campbell Island; this animal presumed by him to have pelage intermediate between hair and fur. The sea lions breeding in waters on both sides of the Tasman Sea have in fact been listed under many generic names, including *Otaria, Neophoca, Gypsophoca, Euotaria, Zalophus, Arctocephalus, Eumetopias,* and *Phocarctos.* Sivertsen (1954, pp. 28–31) assigned them all to one species, *Neophoca cinerea.* Sivertsen's study material included the skulls of 13 specimens (6 adult males) of "*Phocarctos hookeri*" from Campbell Island and 9 specimens (including 2 adult males) of "*Neophoca cinerea*" from Australia. King (in Rand, 1956*b*, p. 6) regarded *Phocarctos* as a distinct, monotypic genus.

The present writer agrees with Sivertsen that differences between the Australian specimens and the New Zealand specimens are not of generic importance, but retains the specific names *cinerea* and *hookeri* in order to identify the two populations. On the basis of Sivertsen's material, and personal examination of skulls in the British Museum (Natural History), the adult male skull of *Neophoca cinerea* is wider, attains a CBL of 308 mm. and a sagittal crest height of 30 mm; that of *Neophoca hookeri* is narrower and attains 346 and 23 mm. respectively. The ranges of the two species have remained apart in historic times by at least 1,930 km. (*i.e.,* New Zealand to southeastern Australia).

Neophoca has been listed as recently as 1951 from Japan (Ellerman and Morrison-Scott, p. 324). The writer has seen no specimens of sea lions from Japan which could not be referred either to *Zalophus* or *Eumetopias.* As presently understood, *Neophoca* is confined to the Southern Hemisphere.

Neophoca cinerea (Péron) 1816. (Australian sea lion)

Otaria cinerea Péron, 1816, vol. 2, p. 54 (*nomen nudum*), p. 77. Kangaroo Island, South Australia. Allen at first (1880, p. 215) considered the name indeterminable but later changed his mind.

"*Otaria cinerea,* Péron," Quoy and Gaimard, 1830, vol. 1, p. 89; atlas, 1833, pls. 12 and 15. Western Port, Bass Strait, South Australia. Good description of an eared seal but quite certainly *Arctocephalus* rather than *Neophoca.* The young were said to be "toute noir" (p. 94) and the adult skins were said to be taken by whalers. The types, a skin and a skull, are now in the Muséum National d'Histoire Naturelle, Paris.

Arctocephalus lobatus Gray, 1828, p. 1, pl. 4, figs. 2 (side view of skull) and 2*a* (dorsal view of skull). Houtman's Abrolhos (= Houtman Rocks), southwestern Australia. This species is the genotype of *Neophoca.* There is in the British Museum (Natural History), a specimen labeled as the type of *A. lobatus.* It is no. 1844.3.19.2, original no. 337*a,* front of jaws with all teeth, subadult male; listed by Gray (1874, p. 83).

"*Zalophus cinereus* (Péron)," Allen, 1892, p. 372. Name "*cinereus*" accepted by Allen; the species held to range throughout New Zealand and Australia.

Neophoca cinerea, Iredale and Troughton, 1934, p. 89 (with synonyms).

Neophoca cinerea, Sivertsen, 1953*b* (*nomen nudum*); 1954, pp. 6, 28–31. The genus *Phocarctos* and its single species *hookeri* placed under *N. cinerea.*

TYPE. None. Species based on Péron's description (1816, p. 77) of "une nouvelle espèce du genre Otarie . . . *Otaria cinerea* N." from "l'île Decrès" (= Kangaroo Island, near Adelaide, South Australia). Sea lions were killed here by the ship's crew for food, oil, and leather in January 1803. Apparently one specimen, of a young animal 33 inches long, was taken to France (Allen, 1880, p. 203, footnote).

RANGE (fig. 2). Now confined to the coast of South Australia. Formerly Port Stephens, 33° S, 152° E, near Newcastle, New South Wales, westward to Pearson Island, 36° S, 134° E, and nearby Neptune Island; Houtman Rocks on the west coast at 28° S, 113° E; southward to Western Port, 38° S, 145° E, near Melbourne (Iredale and Troughton, 1934, p. 89; Sivertsen, 1954, p. 30). Lewis (1942, p. 24) stated that sea lions are no longer seen east of Spencer Gulf, 34° S, 138° E, near Adelaide.

Neophoca hookeri (Gray) 1844. (New Zealand sea lion) New combination.

Arctocephalus hookeri Gray, 1844, pp. 4–5, pls. 14–15; 1847, p. 33. "Falkland Islands and Cape Horn" (actually Auckland Islands, South Pacific Ocean).

Arctocephalus (*Phocarctos*) *hookeri,* Peters, 1866*a,* p. 269. Subgenus *Phocarctos* erected; type (first-listed species) "*Otaria Hookeri. Arctocephalus Hookeri* Gray."

Phocarctos hookeri Gray, 1866*a*, p. 234.

Otaria hookeri, Clark, 1873, p. 754 (fig. of skull), p. 755 (fig. of skull). Clark compared skulls from Auckland Islands collected during the French Expedition of 1837–40 (see Jacquinot) with type specimens of

A. hookeri; showed clearly that the type locality must have been Auckland Islands rather than the "Falkland Islands and Cape Horn" of Gray (1844, p. 5).

Phocarctos hookeri Gray, 1874, p. 29, p. 30 (fig. 15), pl. 20; 1875, p. 11. The latter citation is to a description (published in the year of Gray's death) in which Gray erroneously cites himself as author of the name *Phocarctos*, and persists in showing the type locality of *P. hookeri* as Falkland Islands and Cape Horn.

Neophoca cinerea (part), Sivertsen, 1953; 1954, pp. 6, 28–31.

Phocarctos hookeri, King (in Rand, 1956*b*, p. 6).

TYPES. *Syntypes*: Originally two stuffed skins with their skeletons and the imperfect skulls of two other individuals, collected during the voyage of the ships *Erebus* and *Terror*, 1839–43, under the command of James Clark Ross, probably from the Auckland Islands south of New Zealand; deposited in the British Museum. The present writer has not verified the existence of all the syntypes nor attempted to clear up the confusion in catalogue numbers, sex, and concordance of skin with skull. No type was designated by Gray, although he showed in his 1874 hand list (pp. 29–30) which specimen was the basis of his 1844 figures (pls. 14 and 15). Thus, the principal type is adult female, British Museum (Natural History), stuffed skin no. (?), plate 14; skeleton, no. 1843.11.25.2, original no. 336*a* (or 336*c*?) plate 15.

RANGE (fig. 2). *General*: New Zealand region, breeding between latitudes 51° and 53° S, straggling to 47° S. *Campbell Islands*: Two skulls collected (Sivertsen, 1954, p. 30). *Auckland Islands*: Breeding at Carnley Harbour and Enderby Island; height of pupping season last week of December; nonbreeders "on other islands of the New Zealand Subantarctica" (Turbott, 1952, pp. 198, 211). *Macquarie Islands*: Fur seals (and probably sea lions if ever here) were exterminated by sealers and feral dogs in the mid-nineteenth century (*Antarctic Pilot*, 1948, p. 321). Doutch (1952) did not see any sea lions in a year spent here, 1950–51. G. L. O'Halloran (Secretary for Marine, New Zealand, *in lit.*) regarded this island as "probably south of the species' normal range." *New Zealand*: "Only an occasional straggler to the coast of New Zealand proper and then principally on the southern part of the South Island" (O'Halloran).

Subfamily ARCTOCEPHALINAE von Boetticher, 1934

Arctocephalinae von Boetticher, 1934, p. 359. (fur seals)

Arctocephalina Gray, 1837, p. 582. Including "*Arctocephalus* and *Otaria*" (=family Otariidae). See discussion under Otariinae, p. 53.

Genus **ARCTOCEPHALUS** É. Geoffroy Saint-Hilaire and F. Cuvier, 1826

Arctocephalus É. Geoffroy Saint-Hilaire and F. Cuvier, 1826, p. 553. (southern fur seals)

TYPE. *"Arctocephalus ursinus,"* based on *"Ursus marinus,* Steller," a sea bear thought by the authors (p. 554) to breed in the Aleutian Islands, Falkland Islands, and Cape of Good Hope. Allen (1905, p. 121) has shown, however, that the "type was not *Phoca ursina* Linn., but the Fur Seal of the Cape of Good Hope," *Phoca pusilla* Schreber as presently regarded.

REMARKS. Both root nouns of the word *"Arctocephalus"* are feminine, yet all reviewers—guided by the terminal *"us"*—have treated the word as masculine. Little would be gained in departing from accepted usage.

In the opinion of the writer the southern fur seals include six species: *pusillus, forsteri, doriferus, gazella, australis,* and *philippii.* Where large populations have persisted, particularly near the mainlands of South Africa, South America, Australia, and New Zealand, the task of identifying by species the southern fur seals is not particularly complicated. But where only small, oceanic populations remain alive today, or are represented by museum relics from the nineteenth century, the task is almost impossible. In the following discussion the geographical distribution of species is arranged after the patterns suggested by Allen (1942), Sivertsen (1954), King (1954), and Rand (1956*b*), but does not agree entirely with any one of them. Excluding the species *gazella* and *phillippii* (which see), the species *pusillus, forsteri, doriferus,* and *australis* can be separated on the basis of differences in the adult male skull. Sivertsen (1954, figs. 34–45) has plotted 12 ratios which suggest these differences. Sivertsen's proposal to revive the genus *Arctophoca* is discussed under *Arctocephalus australis,* p. 74.

Successful mating between *Arctocephalus* and *Zalophus* was mentioned on p. 59.

Arctocephalus pusillus (Schreber) 1776. (South African fur seal)

Phoca pusilla Schreber, 1776. Cape of Good Hope. Schreber's earlier volumes were undated. The name *Phoca pusilla* appears in Theil 3, pp. 315 and 584, and as caption to col. pl. 86, separately bound. J. A. Allen (1880, p. 194) gave the date of publication as 1776; G. M. Allen (1939, p. 247) gave 1776, 1777, and 1775, respectively, for the three appearances of the name.

Arctocephalus pusillus, Peters, 1877, p. 506.

Phoca antarctica Thunberg, 1811, p. 321 (*nomen nudum*). "Mare australe" (= Cape of Good Hope).

TYPE. None. Species based on "Petit phoque. Buff. 13, p. 341, tab. 53" (Schreber, 1776, p. 314). Buffon's illustration, copied by Schreber, is a rather good picture of a young fur seal in its first, almost black, pelage. The type locality was given by Schreber (p. 315) as "in den levantischen [= Mediterranean], und nach . . . Buffon, im indischen Meere" but is assumed to be South Africa.

RANGE (fig. 4). According to an authority on the species (Rand, 1956*b*,

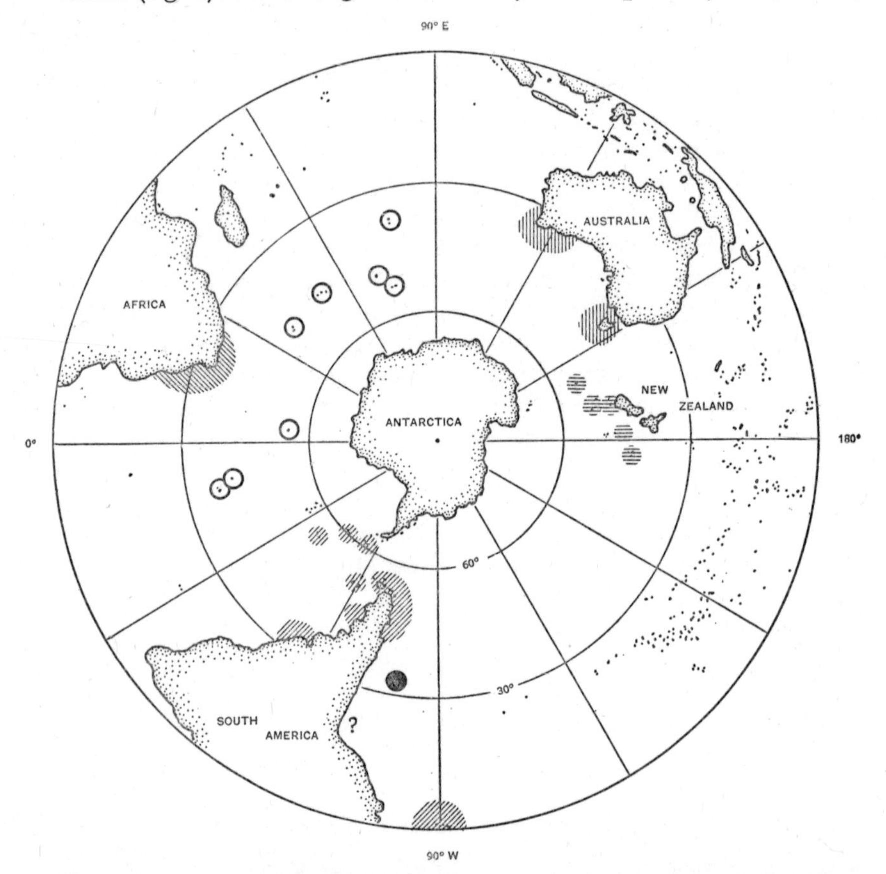

Fig. 4. Ranges of southern fur seals in the Southern Hemisphere: (\\\\) South African fur seal *Arctocephalus pusillus*. (≡) New Zealand fur seal *Arctocephalus forsteri*. (|||) Australian fur seal *Arctocephalus doriferus*. (O) Kerguelen fur seal *Arctocephalus gazella*. (///) South American fur seal: *Arctocephalus australis australis*, Falkland Islands and Falkland Islands Dependencies; *Arctocephalus australis gracilis*, South America; *Arctocephalus australis galapagoensis*, Galapagos Islands (exterminated south of equator). (●) Philippi fur seal (part) *Arctocephalus philippii philippii* (exterminated from type locality and possibly elsewhere). The status of fur seals on the coast of Peru and northern Chile is under investigation.

p. 7 and maps, pp. 34–35) "the Cape fur seal is a coastal species with a rather limited range. Owing to its peculiar breeding habits, distant migrations are not performed and as far as is known the bulk of the population is confined to a comparatively short distance from the land (100 miles or so) . . . The influence of the Benguela current (temperature 10° to 20° C.) is felt over most of the range." There are 17 active populations and at least 10 abandoned rookery sites along the southern coast from South-West Africa (Cape Cross, 22° S) to Cape Province (Bird Island, 33° 51′ S). Whereas the teeth in strongly migratory *Callorhinus* are marked by annual growth ridges (Scheffer, 1950*c*), those in *Arctocephalus pusillus* are not (Rand, 1956*b*, p. 12). The fairly uniform rate of deposition of dentin throughout the year in *A. pusillus* thus seems to be correlated in this case with a resident, rather than a migratory, mode of life.

Its name "*pusillus*" notwithstanding, the South African fur seal may be the largest of all fur seals (Arctocephalinae). Old males attain a weight of 295 kg. (651 lbs.). It is the most widely known of all species of *Arctocephalus*, having been exploited commercially for over 300 years. It is often displayed in European zoos. About 30,000 yearlings of both sexes are killed each winter; about 6,000 bulls each summer; the practice of killing in summer is gradually being discouraged by management agencies (Rand, 1956*b*, p. 32). Cross (1928, p. 39) reported an average annual take of 13,874 skins during the 10-year period 1918–27.

Arctocephalus forsteri (Lesson) 1828. (New Zealand fur seal)

Otaria forsteri Lesson, 1828, p. 421. Dusky Sound, New Zealand.
Arctocephalus forsteri, Gray, 1871, p. 25.

TYPE. None. Species based on George Forster's account of "sea-bears . . . *Phoca ursina* Linnaeus" killed by the ship's crew on several occasions in Dusky Sound, extreme southwestern tip of South Island, New Zealand, in April, 1773, during Cook's second voyage (Forster, 1777, pp. 135 ff.). Forster gave Buffon a sketch and description which Buffon (1782) used in his account of "l'ours-marin," likewise identified with the *Phoca ursina* of Linnaeus. The sea bear of Forster was mentioned by several authors after Buffon, though not until 1828 was it formally described under the name it now carries. Lesson originally included within the range of *forsteri* "l'archipel de Pierre-le-Grand" (= Peter I Island), Antipodes Islands, southern New Zealand, southern Australia, and Tasmania. He also included Tristan da Cunha and certain places along the coast of South America now regarded as being within the ranges of other species than *A. forsteri.* The type locality by restriction is Dusky Sound.

RANGE (fig. 4). Southern New Zealand and nearby subantarctic islands as far south as 55° S (Allen, 1942, pp. 430–31; Turbott, 1952, pp. 198,

212; Gwynn, 1953*b*, p. 19). Certain of the Chatham Islands; the Bounty Islands; rocks and islets surrounding Stewart Island, Snares Islands, and Solander Island; Campbell Island; Macquarie Islands (about 200 non-breeders haul out regularly each summer). Turbott stated that recent visitors to Antipodes Islands and Auckland Islands saw no sign of fur seals. He gave the height of the pupping season in New Zealand waters as early December. Rand (1956*b*, map p. 35) included both of the main islands of New Zealand within the range of *A. forsteri.*

W. R. B. Oliver (*in lit.*, 1948) wrote that "in New Zealand we have the problem of preserving our fur seals, always having to keep on guard lest traders should get a license to kill some. This actually happened recently and 11,000 males, females, and young were killed. I think, however, on account of the public protest that followed, that licenses will not be issued for some time." Falla (1953) estimated the total population of the New Zealand fur seal at "considerably less than 50,000 animals, scattered along two or three hundred miles of coastline, and islands 100 miles or so apart."

REMARKS. Allen (1942, pp. 429–30) listed *A. forsteri* (Lesson), *A. doriferus* Wood Jones, and *A. tasmanicus* Scott and Lord, stating that "whether these are really separate races or species, or whether the characters represent merely individual variations, seems as yet to be uncertain." Sivertsen (1954, pp. 45–48) regarded *A. doriferus* and *A. tasmanicus* as indistinguishable from *A. forsteri.* However "King (in ms.) says that the skulls of *A. doriferus* and *A. forsteri* in the British Museum collection are distinctly different" (Rand, 1956*b*, p. 6). On the basis of Miss King's judgment, and considering that New Zealand and southwestern Australia are 1,930 km. apart, the present writer feels that the specific distinction between *forsteri* and *doriferus* should be upheld. The relation of "*tasmanicus*" to *doriferus* will be discussed below.

Arctocephalus doriferus Wood Jones, 1925. (Australian fur seal)

Arctocephalus doriferus Wood Jones, 1925, p. 12. South Australia.

Arctocephalus tasmanicus Scott and Lord, 1926, pp. 187 and 188 (*nomen nudum*), p. 189, pls. 16–21; spelled *tasmanicum* on p. 192. Tasmania.

Gypsophoca tropicalis Iredale and Troughton, 1934, p. 89. "Probably Tasmania."

Arctocephalus forsteri (part), Sivertsen, 1954, pp. 45–48. Dusky Sound, New Zealand.

TYPE. None. Species based on "the small fur seal of the coasts of South Australia" (Wood Jones, 1925, p. 11). Wood Jones gave a list of eleven synonyms, starting with "the small fur seal of Kangaroo Island, Péron [1816, vol. 2, p. 118]" and following with ten formal names, all variations of *cinereus* or *forsteri.*

RANGE (fig. 4). *General*: Southern Australia and Tasmania, between 25° and 44° S, 148° and 113° E. *Victoria*: "There are about fourteen colonies of Fur-Seals in Bass Strait, the four principal ones being on the Victorian coast and the balance in Tasmanian waters. Our [Victorian] four colonies are at Julia Percy Island (off Port Fairy), the Seal Rocks (off Phillip Island), the Glennies (off Wilson's Promontory), and the Skerries, a group of rocks off the mouth of the Wingan River, about 20 miles west of Mallacoota" (Lewis, 1942, p. 24). The Seal Rocks may have 3,000 to 5,000 seals at the height of the breeding season. *South Australia*: Kangaroo Island (fur seals exterminated in late 1800's), Nuyts Archipelago and Investigator Group (exterminated?), Inner Casuarina Island (at least until 1912). *Western Australia*: Seals are resident among the Recherche Archipelago and perhaps off Cape Leeuwin and Houtman Rocks; sporadically as far north as Shark Bay (Allen, 1942, pp. 430–31; Troughton, 1951, p. 252). The latitude of Shark Bay, 25° S, may represent the northern limit of the range, both of the Australian and the New Zealand fur seals. *Tasmania*: At the southern limit (44° S) there are about 1,000 fur seals on Pedra Branca (= Pedra Blanca), 26 km. south of Tasmania (Fowler, 1947). *Limital range*: Fur seals have been recorded sporadically from lakes and streams in Victoria and New South Wales as far as 80 km. from the sea (Troughton, 1951, p. 251).

REMARKS. Wood Jones (1925, p. 15) compared *doriferus* with *forsteri*. Excluding certain trivial features of color, and length of claws and claw flaps, his comparisons were: *A. doriferus*, length of male about 6 ft., female about 5 ft., CBL of male to about 250 mm., of female to about 200 mm.; front series of cheek teeth with two secondary cusps. *A. forsteri*, length of male 6–7 ft., female 5 to 6.5 ft.; CBL of both sexes to about 230 mm.; cheek teeth with single secondary (anterior) cusp only.

It is doubtful that "*tasmanicus*," whose described range is said to be bounded on the east and west by that of *doriferus*, deserves recognition. The description of "*tasmanicus*" is far from definitive. Thus: "It would appear, therefore, that our most common eared seal [named *tasmanicus* in a paper read 4 months later] is rather larger than . . . *Arctocephalus doriferus*, that its nasal and crest osteological data do not quite agree, but these are minor matters in a way" (Scott and Lord, 1926, p. 78). And 4 months later (1926, pp. 192–94) comparisons were made with respect to nasal bones, postorbital processes, length of molar tooth-row, and CBL, between "*tasmanicus*" on the one hand and "*Arctocephalus cinereus* of Wood Jones" (not of Péron). But Wood Jones (1925, p. 15) had included "*cinereus*" as well as "*tasmanicus*" in the Tasmanian fauna! In summary, the authors of "*A. tasmanicus*" apparently did no more than point out slight differences in skulls of three samples of fur seals, all with ranges overlapping in Tasmania.

But Ellis Le G. Troughton of the Australian Museum has kindly written (*in lit.*, 29 July 1957), "I can definitely confirm the validity of *tasmanica* as a large-skulled species, as distinct from *dorifera* . . . There is in our collection a skull of a male [*A. doriferus*] from Bass Strait which was 14 years in Taronga Park Zoological Gardens (Sydney). It does not nearly approach the size and age characteristics of adult *tasmanica*." The fur-seal populations of Australia and Tasmania are said to be increasing under wise management. Zoologists may eventually be able to obtain samples large enough to prove or deny the existence of local races or species.

Arctocephalus gazella (Peters) 1875. (Kerguelen fur seal)

Arctophoca gazella Peters, 1875, p. 393 (*nomen nudum*), p. 396. By restriction, Kerguelen Islands, southern Indian Ocean.

Otaria (*Arctophoca*) *elegans* Peters, 1876, p. 316. Saint Paul Island or (?) Amsterdam Island.

Arctocephalus gazella (Peters) 1877, p. 507.

Arctocephalus gazella, Allen, 1892, p. 375. Essentially as now regarded.

Arctocephalus (*Arctocephalus*) *gazella,* Trouessart, 1897, p. 374; 1904, p. 281. *A. elegans* regarded as a variety of *A. forsteri.*

Arctocephalus gazella, Allen, 1942, p. 432. *A. elegans* in synonymy.

TYPE. *Syntypes*: (1) Skin, skeleton, and certain soft parts of a young female, total length 83 cm., from Kerguelen Islands, collected in 1874 by Dr. (Theophil?) Studer, Naturforscher on S. M. S. *Gazelle* of the German Transit-of-Venus Expedition, 1874–76. The *Gazelle* wintered at Anse Betsy (49° 09′ S, 70° 11′ E). (2) Skin only, adult, total length 177 cm., judged to be male. The following year, Peters learned that the male skin had actually come from the St. Paul Island–Amsterdam Island group. He reexamined it and made it the type of *Otaria* (*Arctophoca*) *elegans*; whereupon Allen in 1892 returned the species *elegans* to *gazella*!

RANGE (fig. 4). Remote, subantarctic islands in the southern Indian Ocean and southern Atlantic Ocean between 37° S (Tristan da Cunha) and 54° S (Bouvet Island), and between 78° E (St. Paul Island) and 13° W (Tristan). *Saint Paul Island and Amsterdam Island*: The identity of the fur seal here is uncertain. Clark (1875, pp. 673–74) placed four skulls under *Otaria* (= *Arctocephalus*) *forsteri*; King (in Rand, 1956*a*, p. 68) also concluded that "skulls from St. Paul Island have a closer resemblance to those of *A. forsteri* than of *A. gazella*." The present writer has tentatively placed the St. Paul–Amsterdam seals under *A. gazella,* following the lead of Allen (1942, p. 433), Sivertsen (1954, p. 45), and Rand (1956*b*, p. 35). On the basis of his visit in 1955–56, Paulian (1956, p. 5) estimated 1,500 fur seals on these islands. In a later paper (1957*a*) he suggested that *A.*

gazella may in fact consist of two geographic races or subspecies, one south of the "antarctic convergence" and one north. In this part of the western Indian Ocean and eastern Atlantic the convergence lies roughly along latitude 48° S. Seals on the islands of Heard, Kerguelen, and Bouvet seem to fall into one anatomical group and seals on the islands of St. Paul, Amsterdam, Crozet, Marion, Tristan, and Gough into another. A systematic study of *gazella*-like skulls is being pursued by Miss Judith King. *Heard Island and McDonald Islands*: 40 sealers were on Heard in 1874, distributed in parties along the coast (*Antarctic Pilot*, 1948, p. 290). Today "Heard is visited by bands [the largest seen, 31 animals] of fur seals in the late summer and autumn, at a time when their home rookeries are occupied by the breeding animals and their pups. Unless there is some undiscovered breeding place on the McDonald Islands there is little doubt that the fur seals seen at Heard Island are bands wandering south from Kerguelen" (Gwynn, 1953*b*, pp. 9–10). *Kerguelen Islands*: The type locality; 3,000 seals killed here in 1880 (Allen, 1942, p. 433); adult male captured recently (Paulian, 1952); small numbers seen throughout the summer (Angot, 1954, p. 10). *Crozet Islands*: Shown within range of *A. gazella* (Sivertsen, 1954); harems seen on 8 December 1949 by de la Rüe (1950). *Prince Edward Islands*: Prince Edward Island and Marion Island, its neighbor 16 km. away; several hundred seals (Rand, 1956*a*). *Bouvet Island*: 37 specimens collected on the Norwegian (Christensen) Antarctic Expedition, 1928. On nearby Lars Islet, 1,000–1,200 seals were counted (Sivertsen, 1954, pp. 6–7, 62). *Tristan da Cunha group and Gough Island*: Identity of the fur seal here has been questioned. It was tentatively placed by Elliott (1953) and Swales (in Holdgate *et al.* 1956, p. 236), with *pusillus*; later with *gazella* (see *errata, Polar Record,* 1957, vol. 8, p. 491); by Sivertsen (1954, p. 49) with either *pusillus* or *gazella*; by Rand (1956*a*) "related to *A. gazella*." Tristan Island (occasional; Elliott; Swales). Inaccessible Island (Sivertsen obtained a seal flipper in 1928–29; Holdgate estimated 100–200 seals in 1955). Nightingale Island (at least 90 seals in November 1951, presumably from breeding caves on Inaccessible Island; Elliott); Gough Island, 386 km. south-southeast of Tristan da Cunha, more than 2,400 km. from the nearest continent (more than 10,000 fur seals were counted, mostly on the west beaches, in 1955; Swales).

Within the entire range of *A. gazella* as given here, but excluding the Tristan-Gough group, Paulian (1956, p. 5) estimated 3,000 to 4,000 fur seals living today.

REMARKS. Sivertsen (1954, pp. 49–51, 74) has shown that *A. gazella* can be separated from other species of *Arctocephalus* by inspection of its postcanine teeth. (He did not use *A. philippii* in his comparison, regarding it as a member of the genus *Arctophoca.*) The teeth in *gazella* are remarkably slender, widely spaced, and lacking in secondary cusps. Sivertsen applied the formula: (sum of antero-posterior diameter of 6 upper cheek teeth) ÷ (distance from posterior of canine to posterior of 6th cheek tooth) × 100. The result was under 56 percent for both males and females of *gazella* as against over 68 percent for males and females of the other species of *Arctocephalus*. In adult males of *gazella* the gap between the 3d and 4th upper molars was over 1 percent of the CBL; in other species less than 1 percent.

Arctocephalus australis (Zimmermann) 1783. (South American fur seal)

Phoca australis Zimmermann, 1783, vol. 3, p. 276. Falkland Islands, South Atlantic Ocean.

Arctocephalus australis, Allen, 1880, pp. 193, 210.

TYPE. *Holotype*: A seal in the "Museum of the Royal Society . . . sent of late years from the Falkland Islands" (Pennant, 1781, vol. 2, p. 521). The length was said to be 4 feet; teeth, skin, and appendages were described. Zimmermann's distribution map showed *Phoca australis* on the west coast of South America below 40° S, between the Falkland Islands and Cape Horn, and west of New Zealand in the Tasman Sea. The type specimen was apparently lost; it was not listed by Flower (1884, p. 191).

RANGE (fig. 4). From Brazil to Strait of Magellan and northward to southern Peru, Islas Juan Fernández (?), Galapagos Islands, Falkland Islands, and Falkland Islands Dependencies. See further under subspecies.

REMARKS. Sivertsen (1954) and King (1954), studying approximately the same materials, came independently to the conclusion that there are two principal kinds of resident fur seals in the waters of South America and Mexico. The "*australis*" kind has a broad skull; inhabits the mainland of South America, the Falklands, Galapagos, and certain smaller islands. The "*philippii*" kind has a narrow skull and is known only from certain small Chilean islands and—hundreds of kilometers to the northward—the coasts and islands of Baja California and vicinity. King referred the two kinds to *Arctocephalus australis* and *A. philippi,* respectively. Sivertsen referred the broad-skull seals to *A. australis* and *A. galapagoensis*. For the narrow-skull seals he revived the generic name *Arctophoca* Peters (1866*a*, p. 276) and used the combinations *Arctophoca philippii* and *Arctophoca townsendi* for the Chilean and Mexican kinds, respectively.

The present writer feels that distribution and variation in these southern fur seals can best be expressed by use of subspecific names within the species *australis* and *philippii,* both within the genus *Arctocephalus*. *Arctocephalus,* in contrast with the related fur seal *Callorhinus,* is less migratory. Its members (except the ice-front populations) tend to remain the year around in the vicinity of the breeding beaches. Where local populations were exterminated in the nineteenth century, as for example on the California islands, new colonies have failed to reappear. Thus, a few hundred seals on Isla de Guadalupe represent the only surviving individuals of *Arctocephalus* in the Northern Hemisphere. Where, then, surviving populations demonstrate both geographical isolation and differences in skull proportions (though not perhaps rising to the level of statistical significance) the situation is best described by assignment of subspecific names.

Sivertsen's proposal to recognize two genera of southern fur seals, while straightforward, falls somewhat short of convincing. First, the "*Arctophoca*" material is scanty. In addition to a drawing of a skull, only 4 complete or nearly complete skulls were available to him: a male and a female from Chile and a male and a female from Mexico (1954, p. 44). Second, none of the specimens is very old. The oldest has a suture age of 25 out of a possible 36 (1954, p. 15). Third, the only important difference between "*Arctophoca*" and other fur seals seems to be its narrower skull. That is, in a comparison of 12 skull ratios among "*Arctophoca*," *Callorhinus*, and *Arctocephalus*, 5 ratios in "*Arctophoca*" are distinct, though all 5 illustrate narrowness. The comparisons are based on only 2 skulls of "*Arctophoca*" plus measurements of a third taken from Peters' drawing. Having some first-hand knowledge of the great variability in otariids, the present writer does not feel that the differences shown by Sivertsen are important at the generic level. Fourth—and this is a weak argument—it does not seem reasonable that a distinct genus would be represented in the peculiar geographic pattern suggested for "*Arctophoca*," namely: Chilean islands but *not* on the Galapagos Islands; and again on the Mexican islands—nowhere else in the world.

The present writer feels that, were only skull characters to be weighed, a split at the subgeneric level between the "*australis*" kind and the "*philippii*" kind might be warranted. Peters, in fact, made the split at that level and Allen (1905, p. 124) agreed. But considering the geographical relationships of the "*philippii*" kind to all other southern fur seals, such distinction does not seem to be warranted.

Arctocephalus australis australis (Zimmermann) 1783. (See origin of name under species.)

RANGE. *Falkland Islands*: Laws (1953*a*) estimated over 20,000 seals breeding in the summer of 1951 on 10 or more beaches. *Falkland Islands Dependencies*: Included under *A. a. australis* for convenience, though not certainly of this race, are fur seals of the Dependencies. Laws (1953*a*) estimated 5,000 seals here. South Georgia (1,000–3,000 on Bird Island), sporadically to South Orkney Islands and South Shetland Islands. Matthews (1952, p. 87) stated that "small rookeries on the north-western islets and in the bays of the south-east coast [of South Georgia] are increasing in size." Fur seals, once present in the South Sandwich Islands, were not seen here by visitors in the 1930's (*Antarctic Pilot*, 1948, p. 122).

REMARKS. The seals of the South Shetlands and South Georgia were named "*Arctocephalus shetlandii*" by Brass (1911, p. 665). No type was designated, no measurements were given; the pelt was simply said to be considerably better than that of the Alaska seal; large, very finely and thickly haired.

Arctocephalus australis gracilis Nehring, 1887.

"*Arctocephalus falclandicus* Shaw (*Arctophoca falclandica* Burm) . . . [var.] *gracilis*" Nehring, 1887*a*, pp. 92–93. Lagoa Tramandaí, Rio Grande do Sul, Brazil. Nehring also suggested (p. 92), but with less conviction, the combination "*Arctocephalus (Arctophoca) gracilis.*"

Arctocephalus falclandicus var. *gracilis* Nehring, 1887*b*, p. 142.

Arctocephalus australis gracilis, King, 1954, p. 320.

TYPES. *Syntypes*: Three skulls only, in good condition, young male and two young females, Zoologisches Museum, Humboldt-Universität zu Berlin nos. 4315, 4316, and 4317, respectively; from the coast of southern Brazil (Estado de Rio Grande do Sul, Lagoa Tramandaí, 29° 55′ S); before 1887; collected by Theodor Bischoff, a teacher from Mundo Novo. Bischoff also provided a description of the skin of a young seal from the same place (actually 6 or 7 km. inside Rio Tramandaí) and the skin of a 7-foot adult.

RANGE. *General*: Brazil to Strait of Magellan and northward to southern Peru. *Brazil*: Göldi (in Nehring, 1887*c*, p. 207) reported the capture of fur seals at Ponta Negra (22° 56′ S). Cabrera and Yepes (1940, p. 180) reported them as far north as the mouth of the Rio de Janeiro. Vaz Ferreira (*in lit.*) visited Recife dos Torres, between Santa Catalina and Rio Grande do Sul, in April 1953. On this low islet he saw a few fur seals. In the Museo Julio de Castilhas, Rio Grande do Sul, he saw mounted specimens of young fur seals from the islet. He believes that Recife dos Torres may represent the northern limit of breeding. *Uruguay*: Vaz Ferreira (1956*a, b*) found the only remaining breeding colonies on six islands, as follows: Islas de Castillo Grande (Isla del Marco), Islas de Torres (Isla Rasa, Islote de Torres, Isla Encantada), Islas de Lobos (Isla de Lobos, Islote de Lobos). He had earlier (1950, p. 155) estimated the population on Isla de Lobos at 56,000. *Argentina*: Santiago Carrara (1952, appendix p. 4) estimated 290 fur seals on Isla Escondida but was unable to survey the rookeries of Isla de los Estados. Later (1954) he estimated 2,700 fur seals in the waters of Argentina, as follows: Isla Escondida, 400; Isla de los Estados (Bahía San Antonio, 100; Bahía Flinders, 2,200). The Argentine fur seals are not as yet commercially exploited. *Pacific coast of South America*: Present distribution along the coasts and islands of the west coast is virtually unknown to the writer. Cabrera and Yepes (1940) stated that the fur seals had been gone for many years from the northern and central coasts of Chile. Milton J. Lobell, for several years a Fish and Wildlife Service fisheries specialist on the west coast, wrote (*in lit.*) that "fur seals . . . are only very rarely encountered in Peruvian waters and then most commonly in the extreme south . . . In Chile, however, fur seals still seem to be fairly common. Certainly they do exist in the [Archipiélago de los Chonos] southward, for I have seen individuals from time to time." Yañez (1948, p. 110) stated that the fur seal once lived along the entire coast of Chile but is now confined to Tierra del Fuego.

REMARKS. King (1954, p. 312) has shown that *A. australis* on the South American mainland, including nearby islands, is slightly smaller than the fur seal of the Falkland Islands. For example, the ratio ZW/CBL is significantly different in samples of the two populations. The present writer feels that, in view of the demonstrated differences in their skull characters and their geographic isolation, the two populations need separate names. Therefore "*gracilis*" as used by Nehring for the mainland race is restored.

Arctocephalus australis galapagoensis Heller, 1904.

Arctocephalus galapagoensis Heller, 1904, p. 245. Isla Wenman, Galapagos Islands.

Arctocephalus galapagoensis, Allen, 1905, pp. 123, 134. Placed in synonymy of *A. philippii.*

Arctocephalus australis galapagoensis, King, 1954, p. 320.

TYPES. The description was based on 200 skins taken among the islands for commercial use, body measurements of 5 adult males and 2 adult females, an unspecified number of immature skulls, and 3 adult skulls. One of the latter is the holotype and two are regarded by Mayer (1949*a*, p. 32) as paratypes. *Holotype*: Skull only, old adult male, Stanford University Natural History Museum no. 2812 (not 2480), original no. 10982; collected by Edmund Heller and Robert E. Snodgrass in 1898 or 1899 on Isla Wenman, Galapagos Islands. (Mayer, *op. cit.*, gave the date as "1888–1889." While Heller gave no date, he specified on page 246, "Wenman Island (Hopkins Expedition)," which could have meant only the expedition of 1898–99.) *Paratypes*: (1) Skull only, nearly adult male, S.U.N.H.M. no. 4442 (not 2481), original no. 1469, collected by Heller and Snodgrass on Isla Wenman in January 1899; (2) skull only, nearly adult female, S.U.N.H.M. no. 4446 (not 2482), original no. 1, collected by Heller and Snodgrass on Isla Wenman in December 1898.

RANGE. Now, sadly, only on Isla Genovesa (= Tower Island), an island about two miles in diameter, to the northeastward of the main archipelago (Banning, 1933; Slevin, 1935).* From the time of their discovery in 1535,

* As the book goes to press, Eibl-Eibesfeldt reports in *Unesco Courier* (11th year, no. 1, p. 20), "On James Island (also known as San Salvador . . .) we discovered a large colony of fur seals which were said to have become almost extinct."

the Galapagos Islands were often visited by whalers and sealers. Capt. Morrell took about 5,000 sealskins here in 1823 (Baur, 1897). During a sealing expedition of 5 to 6 months' duration in 1898–99, a schooner took only 200 skins, chiefly on Isla Wenman, Isla Fernandina (= Narborough Island), and Isla Isabela (= Albemarle Island) (Heller, 1904, p. 246). One seal was seen on Isla Culpepper; seals were reported on Isla Genovesa (= Tower Island) and Isla Pinta (= Abingdon Island). Heller believed that each group was strictly resident. "The seals are now so reduced in number and so scattered that no well-defined rookeries exist . . . the seals being widely scattered and well concealed in holes and crevices" (p. 247). The breeding season was thought (perhaps without foundation) to be indefinite.

REMARKS. King (1954, p. 320) concluded that the fur seals of the Galapagos Islands are probably smaller than those of the mainland. Sivertsen (1954, pp. 48, 52, 53) showed graphically that adult skulls, both male and female, from the Galapagos are considerably shorter than those from the mainland-plus Fallkands. The adult male Galapagos skull is slightly narrower in the interorbital region, higher at the auditory meatus, and stubbier-nosed. The last-named character, expressed as distance from gnathion to interorbital foramen, is especially distinct. The skull of *A. a. galapagoensis* may in fact be the smallest (in both sexes) of any form of *Arctocephalus*. Sivertsen (1954, pp. 49, 72) gave the CBL of full-grown males as 212 mm., of full-grown females as 184 mm. Thus it seems wise to retain the name proposed by Heller for the Galapagos animal, though at the subspecific rather than the specific level.

Arctocephalus philippii (Peters) 1866. (Philippi fur seal)

Otaria (Arctophoca) philippii Peters, 1866*a*, p. 276, pl. 2. Isla Más a Tierra, Islas Juan Fernández, Chile.

Arctocephalus philippii Peters, 1877, p. 507. For other synonyms see Osgood (1943).

TYPE. *Holotype*: Skin and skull of young adult male from Isla Más a Tierra, Islas Juan Fernández, Chile, collected in December 1864 by Rodolfo Amando Philippi. Both skin and skull were lost by fire in World War II (Klaus Zimmermann, Museum für Naturkunde, Berlin, *in lit.*).

RANGE (figs. 4 and 5). Until the nineteenth century, the Philippi fur seal was distributed in two isolated groups off the Pacific coast of the Americas at 30° N (Baja California) and 30° S (Chile). Now apparently restricted to several hundred animals on Isla de Guadalupe, off Baja California.

REMARKS. It has seemed advisable to split *A. philippii* into two races

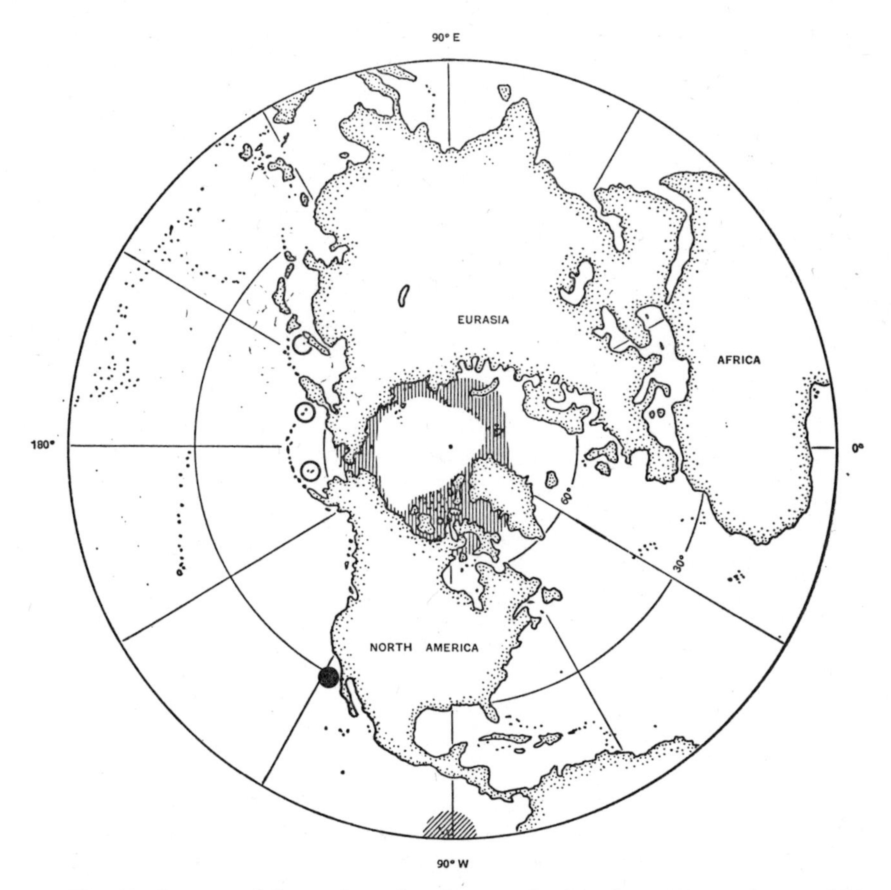

Fig. 5. Ranges of fur seals and walrus in the Northern Hemisphere: (///) South American fur seal (part) *Arctocephalus australis galapagoensis.* (●) Philippi fur seal (part) *Arctocephalus philippii townsendi.* (O) Northern fur seal *Callorhinus ursinus,* breeding range (east to west) Pribilof Islands, Commander Islands, Robben Island; winter range from about 30° N northward to the edge of ice. (|||) Walrus *Odobenus rosmarus rosmarus,* Atlantic-Arctic; *Odobenus rosmarus divergens,* Pacific-Arctic (boundaries of breeding range uncertain).

in order to express the differences found by Sivertsen (1954, p. 42) between the "*philippii*" seals of Mexico and those of Islas Juan Fernández. Sivertsen split the two groups at the species level; King held them in a single species; the present arrangement is thus a compromise. (See discussion under *A. australis,* p. 74.) The skull characters which separate *A. philippii* and *A. australis,* the only other species with which it might be confused, are best seen in King's illustrations (1954, figs. 1 and 2). The skull of *A. philippii* is slenderer and more delicate; the palate is more deeply concave between postcanines 1 and 3. The narrowness of the skull has been expressed by Sivertsen (1954, figs. 9, 11–12, 15–16) in a comparison of *Arctophoca* (= *Arctocephalus philippii*) and *Arctocephalus* (5 species together). Expressed as percent of CBL in adult males, the main differences are:

	A. philippii	Other species of *Arctocephalus*
Interorbital width	< 12 %	> 12 %
Occipital crest to mastoid	< 42 %	> 42 %
Mastoid width	< 50 %	> 51 %
Width at canines	< 20 %	> 19 %
Width of condyles, lower jaw	< 12 %	> 13 %

Carl L. Hubbs (*in lit.*) suggested that the skull of the Guadalupe seal may be distinguished from that of the Chilean subspecies by the narrower diastema between the 4th and 5th upper postcanines of the former (*cf.* King, 1954, figs. 3 and 1).

Arctocephalus philippii philippii (Peters) 1866. (See origin of name under species.)

RANGE. Under one name or another this "Chilean fur seal" is known reliably from only Islas Juan Fernández (Isla Más a Tierra and its outlying neighbor Isla Más Afuera, 800 km. west of Chile). Sivertsen (1954, p. 6) and King (1954, p. 327) agreed on this point; King suggested that *A. philippii* may have occurred also on Archipiélago de los Chonos and the Galapagos Islands. Two or three million fur seals were taken from the Islas Juan Fernández before 1824 (Kellogg, 1942, p. 459). Fifty skins were taken on Isla Más a Tierra in 1898 and sold in London for from 13 to 32 shillings each (Cabrera and Yepes, 1940, p. 181). Nybelin (1931, p. 494) recorded a new cestode genus and species from "*Arctocephalus australis* (Zimm.)" (?) collected on Isla Más a Tierra on 28 January 1917. According to Luke (1953, p. 292) there are no seals on the island today.

The writer cannot determine from the evidence at hand whether *A. philippii* existed (or still exists) along the mainland of South America. He does not believe that at any time two forms of *Arctocephalus* bred on the same island (p. 49).

Arctocephalus philippii townsendi Merriam, 1897. New combination.

Arctocephalus townsendi Merriam, 1897, p. 178. Isla de Guadalupe, Baja California, Mexico.

"California Fur-Seal. *Arctocephalus sp. nov.*" Allen, 1892, p. 373. "Dr. Merriam has recently obtained skulls from the old killing ground on Guadalupe Island which show that it is not only a different, and as yet a probably undescribed species, but that it is referable to the genus *Arctocephalus,* not previously known to occur north of the Equator."

TYPES. Beachworn skulls only, collected by C. H. Townsend (at West Anchorage) on Isla de Guadalupe, Baja California, Mexico, on 22 May 1892. *Holotype*: Skull of adult male lacking mandibles and teeth, U.S. National Museum no. 83617. *Paratypes*: Skull of subadult female with both jaws and teeth, no. 83618; two skulls lacking both the jaws and face, nos. 83619 and 83620.

RANGE. Breeding in 1957 only on Isla de Guadalupe; population 200 to 500 (the writer's observations in 1955 and reports by Hubbs, 1956*a, b*). Sporadically to southern California: 3 adult males on Piedras Blancas in 1938 (Bonnot, Clark, and Hatton, 1938, p. 416; Bartholomew, 1950, p. 176); 1 adult male on San Nicolas Island in 1949 (Bartholomew, 1950).

The primitive range may have included coastal islands from about Farallon Islands, California (37° 44′ N), to Islas San Benito, Mexico (28° 18′ N). One might suppose that *Arctocephalus* did not breed north of Point Conception (34° 27′ N), where sea-water temperatures break sharply. On the other hand, commercial sealing records from the early part of the nineteenth century indicate that large numbers of fur seals were taken on the Farallons; they were taken at a season of the year when other fur seals (*Callorhinus ursinus*) were presumably rare or absent; and they were taken in diminishing numbers. From the evidence, a local population of *Arctocephalus* was being destroyed in that period. "As recorded by Bancroft . . . in July, 1810, Gale was left on the Farallone Islands with seven men. Here he found two other bands of men that had been left a short time before from other ships, apparently of the same ownership and interests as the Albatross. By December, when the Farallone party was communicated with, they had taken 30,000 sealskins. In three years they took as follows: *1810,* 33,740 skins; *1811,* 21,153 skins; *1812,* 18,509 skins; total for Farallones, 73,402 skins . . . From 1812 to 1840 the Russians kept up an establishment at the Farallones as well as at Ross. The object at first was to secure fur seals, 1,200 to 1,500 skins being taken annually for five or six years, though Winship, Gale, Smith, and other Americans had taken the cream of this natural wealth a few years earlier. After 1818 the seals diminished rapidly until only 200 or 300 per year could be caught" (Starks, 1922, pp. 156–58).

King (1954, pp. 327–29) gave the range as Farallon Islands, Santa Barbara Islands (Santa Barbara, San Nicolas, Santa Cruz, Santa Rosa, San Miguel), Ventura, Isla de Guadalupe, and Islas San Benito. Bone fragments in middens in Ventura County, California, suggest either that seals frequented the mainland or (more likely) nearby islets from where their carcasses were carried to the mainland by aborigines. Hanna (1926, p. 9) was instructed to make a search for fur seals "of which it is believed a few still remain about [Islas Revillagigedo, 18°–19° N], particularly Socorro." The present writer has been unable to verify the existence of fur seals, or any pinnipeds, at any time in the past among these subtropical islands. Berdegué (1956, pp. 21, 28), however, placed the islands within the historic range of the fur seal.

Genus **CALLORHINUS** Gray, 1859

Callorhinus Gray, 1859, p. 359.

TYPE. *Phoca ursina* Linnaeus. The specimen examined by Gray was an adult (about 10-year male) skin and broken skull, British Museum (Natural History) no. 1859.1.17.1, original no. 1221*a*, from Bering Strait, purchased from "Franks," figured by Gray (1859, pl. 68; 1874, pl. 19).

REMARKS. "It is said that sea lions sometimes invade fur-seal rookeries in the Commander Islands and violate the females there. The resulting progeny yield very poor fur and care is therefore taken to destroy them on fur-seal farms" (Naumov, 1933, p. 23). The present writer knows of no case of interbreeding between *Callorhinus* and *Eumetopias* on the Pribilof Islands, where the two animals breed side by side and where their behavior has been carefully watched for more than a half-century.

Callorhinus ursinus (Linnaeus) 1758. (northern fur seal)

Phoca ursina Linnaeus, 1758, p. 37. Bering Island, Commander Islands.

Callorhinus ursinus, Gray, 1859, p. 359.

Callorhinus ursinus, Taylor *et al.,* 1955, p. 61. Members of the three main breeding groups regarded as indistinguishable. Synonymy follows:

Pribilof Islands group

Siren cynocephala Walbaum, 1792, p. 560. Pacific Ocean about 53° N, 155° W, south of Kodiak Island, Alaska, where Steller saw the "sea ape" (Stejneger, 1936, pl. 12). From its location Stejneger assumed—rightly no doubt—that the animal was a member of the Pribilof-breeding population.

Arctocephalus californianus Gray, 1866*b*, p. 51. Skin and broken skull of (2-to-4-year) male, British Museum (Natural History) no. 1859.11.5.2, original no. 1320*b*; collected at Monterey, California, by J. H. Gurney.

Callorhinus alascanus Jordan and Clark, 1898, p. 45 (with further description in 1899, pp. 2–4). "Pribilof herd." (Mayer, 1949*a, b,* has chosen a lectotype from material which the original authors probably examined; they did not identify any specimens by number.)

Callotaria alascana, Allen, 1905, p. 123.

Callotaria ursina cynocephala, Stejneger, 1936, p. 285.

Callorhinus ursina cynocephala, Hall, 1940, p. 76.

Callorhinus ursinus cynocephalus, Allen, 1942, p. 444; Scheffer, 1942, p. 43.

Robben Island group

Phoca mimica Tilesius, 1835, p. 715. Bay of Patience (= Taraika Bay), Sakhalin Island (Stejneger, 1936, p. 281).

Callorhinus curilensis Jordan and Clark, 1898, p. 45 (with further description in 1899, pp. 2–4). "Robben Island herd."

Callotaria curilensis, Allen, 1905, p. 123.

Callotaria ursina mimica, Stejneger, 1936, p. 286.

Callorhinus ursinus mimicus, Allen, 1942, p. 442.

TYPE. None. Species based on Steller's description (1751, pp. 346–59) of seals studied on Bering Island, Commander Islands, in the summer of 1742; and on sketches made by Plenisner (and Berckhan?) under Steller's direction sometime in 1742–43, reproduced as pl. 15, fig. 1 (male on a rock) and fig. 2 (female lying on her back) in *De Bestiis marinis;* sketches subsequently lost. Before 1741, Steller had known of the fur seal as an object of commerce in Kamchatka, and on 10 August 1741 he saw the "sea-ape" (*Siren cynocephala* Walbaum) without then or later recognizing it as a fur seal. The type locality given by Linnaeus was "Camschatcae maritimus inter Asiam & Americam proximam, primario in insula Beringii."

In summary, the original name "*ursinus*" is based upon a detailed description of fur seals on one of the Commander Islands; the name "*cynocephalus*" upon a brief description of a lone seal observed at sea 290 km. off the Alaskan coast; the name "*mimicus*" upon a group of seals observed in a bay of the Sea of Okhotsk. None of the three names was attached by its author to a museum specimen; only one ("*ursinus*") was identified by sketch. The name "*ursinus*" has not been challenged from the time it appeared in the 10th edition of *Systema Naturae.* While the names "*cynocephalus*" and "*mimicus*" are not so well founded, they are the first valid names applied to fur seals from the northeastern Pacific Ocean and Sea of Okhotsk, respectively.

RANGE (fig. 5). *In summer and fall:* Breeding on Pribilof Islands, Alaska; Bering Island and Copper Island (= Ostrov Beringa and Ostrov Mednyy), Commander Islands, U.S.S.R.; and Robben Island (= Ostrov Tyuleniy), off Sakhalin, U.S.S.R. Breeding populations apparently no longer exist on the Kuril Islands. However, Dr. Sergei W. Dorofeev told the writer in 1958 that several thousand fur seals, including at least one black pup, had recently been seen on Ostrova Lovushki, 48° 32′ N, 153° 51′ E. *In winter and spring:* Widely dispersed over the North Pacific Ocean, southern Bering Sea, Sea of Japan, and Sea of Okhotsk; as far south as waters off San Diego, California (32° 40′ S), and Japan (Honshu, 30° N). Rarely straggling into the Arctic Ocean. McEwen (1954) reported a fur seal taken in a fresh-water lake in Yukon Territory, 68° 48′ N, 136° 42′ W.

Populations (summer) are estimated at: Pribilof, 600,000 newborn and 1,200,000 older; Commander Islands, 20,000 newborn and 40,000 older; Robben Island, 20,000 newborn and 40,000 older (U.S. Senate, 1957, and other sources). A popular acount, not seen by the writer, would credit Robben Island with 75,000 seals (Anonymous, 1957, p. 63). Kenyon and Wilke (1953) have given a detailed account of the movements of the fur seal, with information on the recovery of tagged specimens.

REMARKS. Taylor *et al.* (1955) concluded that the fur seals of American waters cannot be distinguished from the fur seals of Asian waters on the basis of color, body size, or skull size. They reported evidence of substantial intermingling of marked seals in winter off Japan and very slight intermingling of marked immature animals in summer on the Commander Islands and Robben Island.

Family ODOBENIDAE Allen, 1880

Odobaenidae Allen, 1880, p. ix (*nomen nudum*), p. 5. Emended to Odobenidae by Palmer, 1904, p. 833. (walrus family)

Odobaeninae Orlov, 1931, p. 69. "Die Odobaenidae werden von einigen Forschern . . . als Unterfamilie (Odobaeninae) der Otariidae angesehen."

Genus **ODOBENUS** Brisson, 1762

Odobenus Brisson, 1762, pp. 12, 30.

TYPE. *Odobenus* Brisson (= *Phoca rosmarus* Linnaeus). Hemming (1955) recommended that the International Commission on Zoological Nomenclature use its plenary powers to validate *Odobenus* as against *Rosmarus* Brünnich, 1771, p. 38.

Odobenus rosmarus (Linnaeus) 1758. (walrus)

Phoca rosmarus Linnaeus, 1758, p. 38. Arctic Ocean.

Odobenus rosmarus, Steenstrup, 1860, p. 446.

Trichechus rosmarus, Trouessart, 1897, p. 375; 1904, p. 281.

Odobenus rosmarus, Trouessart, 1905, p. 874. Change of name of family and genus, after Palmer (1904, p. 833).

TYPE. None. The walrus was known in European literature for at least 200 years before Linnaeus. "Habitat intra Zonam arcticam Europae, Asiae, Americae" (Linnaeus, 1758, p. 38).

RANGE (fig. 5). Principally open waters of the Arctic Ocean around the edge of the polar ice; southward in winter; southern limits of regular occurrence about 58° N in Hudson Bay and Bering Sea; southernmost record 42° N, Massachusetts. The circular distribution of the walrus seems to be interrupted—or perhaps one should say constricted—at two places: the central Canadian–Arctic Archipelago and the Severnaya Zemlya Archipelago. The two segments of the population thus demarcated are regarded as subspecifically distinct. The present writer has been able to obtain only meager information on the breeding sites of the walrus in Eurasian waters.

Reports from zoologists in northeastern Canada and Alaska suggest that the future of the walrus is insecure. The naturally slow rate of reproduction of the species, together with wasteful killing by native peoples, has brought diminution both in numbers and in extent of range (Spärck, 1956).

REMARKS. Allen (1880, pp. 156–71) gave a long description, and eleven pairs of sketches, to illustrate the differences between the skulls of "*rosmarus*" (= subspecies *rosmarus*) and "*obesus*" (= subspecies *divergens*). The tusks of *O. r. rosmarus* are shorter, slimmer, straighter (less incurved), and actually *more* divergent than in *O. r. divergens.* Degerbøl (1935, p. 46) concluded from study of specimens that "the principal differences between the two forms of walruses lie in the longer tusks of the Pacific form, its greater facial breadth, and smaller occiput . . . The two forms may be regarded as two separated, but closely related, species." Bobrinskoi (1944, p. 167) stated that in the Atlantic-Arctic race the adult skull width at the level of the canines is 60 to 64 percent of the width between mastoid processes; in the Pacific-Arctic race (with broader snout), 64.5 to 89 percent.

Odobenus rosmarus rosmarus (Linnaeus) 1758. (See origin of name under species.)

RANGE. *Eurasia*: Severnaya Zemlya, Kara Sea, Novaya Zemlya, Franz Josef Land, Barents Sea, mouth of White Sea (occasional), Spitsbergen; Iceland and Greenland (occasional); recorded from Scandinavia, British Isles, and Netherlands. The main Eurasian population stays east of Poluostrov Kanin (45° E). Walruses, up to several hundred at a time, come close to the coast in winter and move gradually northward in summer. *Eastern North America*: Western Greenland (walrus concentrations in the Holsteinsborg district are important to the natives; 400–600 walruses killed and landed annually; total population about 7,000–10,000. Northward to unknown limits, perhaps to north coast of Ellesmere Island, certainly to Barrow Strait and Cornwallis Island (75° N). Westward to about Barrow Strait and Cornwallis Island (95° W), then rare or casual. Southward to Hudson Bay and Hudson Strait; in historic times as far south as Magdalen Islands in Gulf of St. Lawrence (47° N), Sable Island (44° N, where walruses bred), and even to Massachusetts (42° N).

According to Dunbar (1954, p. 14) only one population of walrus in Canada is known certainly to remain in the same locality during winter;

that of the Igloolik region (69° N, 82° W) in northwestern Foxe Basin. Others to the southward (Frobisher Bay, Akpatok Island, Coats Island, and Southampton Island) are transient and their winter distribution is unknown. "There are indications, arising from the dates at which the walrus appear at various places, that the Hudson Bay and Hudson Strait animals, including the Frobisher Bay group, may be parts of a large population which moves into Hudson Strait in the spring and out again in the late fall, wintering perhaps in the neighbourhood of the ice edge in Davis Strait."

Principal authorities on the range of the Atlantic walrus are: Naumov (1933), Saemundsson (1939), Allen (1942), Anderson (1946), Dunbar (1949 and 1954), Ellerman and Morrison-Scott (1951), and Vibe (1956).

Odobenus rosmarus divergens (Illiger) 1815.

Trichechus divergens Illiger, 1815, p. 68. Between Icy Cape and Point Lay, Alaska. Partially described under *Trichechus rosmarus* on p. 64.

Odobenus divergens, Stejneger, 1914, p. 145.

Odobenus rosmarus divergens, Pohle, 1932, p. 78.

TYPE. None. Species based mainly on the observations of Capt. James Cook in July 1779 in Bering Sea and Chukchi Sea (Cook, 1785, vol. 3, chaps. 3–4). Locality stated by Illiger (p. 68) to be "westlichen Nord-Amerikanischen und nahen Ost-Asiatischen Küste, und dem Eise dieser Meere." Allen (1942, p. 469) gave the type locality as "about 35 miles south of Icy Cape [70° 22′ N], Alaska." Allen (1880, pp. 17–18) stated that both Pennant and Shaw had been aware, even before Illiger published, of a difference in the tusks of the Atlantic and Pacific walruses.

RANGE. This Pacific subspecies is believed to integrade with *O. r. rosmarus* in the central Canadian–Arctic Archipelago and in the vicinity of Severnaya Zemlya. The migratory habits of *O. r. divergens* are better understood than those of the Atlantic race, though much awaits to be learned. *Summer*: From Dolphin and Union Strait (69° N, 115° W, rare) westward to west coast of Banks Island, Herschel Island, Point Barrow, Chukchi Sea, East Siberian Sea, and Laptev Sea, though scarce between Chaunskaya Guba (170° E) and Khatangskiy Zaliv (115° E). Southward to Gulf of Anadyr; a few males to Walrus Island (56° N, 161° W) in Bristol Bay. Northward to Wrangell Island (72° N). *Winter*: Bering Sea from Bristol Bay westward to Mus Navarin (179° E). Southward to Pribilof Islands (57° N; one live walrus seen every 5 or 10 years); Commander Islands (none since 1900); Honshu, Japan (41° N, 142° E; a straggler in March 1937).

A map prepared by Brooks (1954, p. 12) shows distribution of the walrus in eastern Bering Sea and adjacent Arctic Ocean. At the southeastern extremity of the range, in Bristol Bay, the walrus is shown as resident the year around. About 1930 a walrus was killed in Port Wells, Prince William Sound (61° N, 148° W) and the mounted head was displayed in "Pioneers' Igloo," Cordova. Another walrus was killed on southern Kodiak Island (57° N) on 15 May 1953 (Will Troyer, *in lit.*). A map prepared by Rass *et al.* (1955, p. 104) showed the concentration points in northwestern Bering Sea; none west of Mus Navarin.

The annual kill of walruses by Alaskan and Siberian natives has been placed at less than 2,000 (Dunbar, 1949, p. 7). Fay (1957, p. 439) has recently expressed his conviction that the total size of the stock is only 40,000–50,000.

Principal authorities on the range of the Pacific walrus are: Kuroda (1938), Barabash-Nikiforov (1938), Bobrinskoi (1944), Anderson (1946), Brooks (1953 and 1954), and Rass *et al.* (1955).

Superfamily PHOCOIDEA Smirnov, 1908

Phocoidea Smirnov, 1908, pp. 1, 36; Gregory and Hellman, 1939, p. 313. See discussion on p. 52 of the two main divisions of the Pinnipedia. (reptigrade seals)

Family PHOCIDAE Brookes, 1828

Phocidae Brookes, 1828, p. 36. "Phocidae. (Brachiodonta)" (= Phocidae as presently recognized). "Brookes, in 1828, was the first to accord to the Earless Seals the rank of a family" (Allen, 1880, p. 413). (phocids)

Phocidae Gray, 1825, p. 340. Including most of the pinnipeds, except walrus, plus sea otter.

REMARKS. Thirteen genera of phocids are recognized; seven of other pinnipeds. Certain phocids have pushed into high latitudes and to the edge of permanent ice, others have persisted in subtropical waters, and a few are relict in lakes. Like all other pinnipeds, the phocids have failed to proliferate in warm seas.

Subfamily PHOCINAE Gill, 1866

Phocinae Gill, 1866*a*, pp. 5, 8 (= Phocinae plus *Monachus*).

Phocinae Trouessart, 1897, p. 382. As presently recognized.

REMARKS. There is no vernacular name for the Phocinae. The group includes the harbor, ringed, grey, ribbon, harp, and bearded seals of northern waters.

Tribe PHOCINI Chapskiy, 1955

Phocini Chapskiy, 1955*a*, p. 164 (diagram), p. 169.

REMARKS. There is no vernacular name for the Phocini. The group includes the harbor, ringed, grey, ribbon, and harp seals. Chapskiy (pp. 164, 170, 188) would further divide the group into subtribes Phocina and Histriophocina.

Genus **PHOCA** Linnaeus, 1758

Phoca Linnaeus, 1758, p. 37.

TYPE. *Phoca vitulina* Linnaeus, by restriction.

Phoca vitulina Linnaeus, 1758. (harbor seal)

Phoca vitulina Linnaeus, 1758, p. 37. European seas.

TYPE. None. The common seal of the shores of middle and northern Europe was known to writers long before Linnaeus. The type locality according to Linnaeus is "in mari Europaeo"; though Thomas (1911, p. 134) would restrict it to "Mari Bothnico et Baltico." Bobrinskoi (1944, p. 175) states that *P. vitulina* enters the Baltic but not the Gulf of Bothnia or Gulf of Finland.

RANGE (fig. 6). The distribution of *P. vitulina* can be represented by a broken circle around the rim of the Arctic Ocean, with four arms extending southward along the shores of Eurasia and North America. The northern limit of the range is perhaps Ellesmere Island (76° N); few individuals reach 70° N. The southern limits are: in the Atlantic Ocean, Portugal (41° N) and North Carolina (35° N); in the Pacific Ocean, northern Baja California (28° N) and China (certainly 37° N, perhaps 32° N). Harbor seals breed in the Baltic but not in other large, saline, or brackish seas of central Europe. Perhaps the largest fresh-water lake which they regularly inhabit is Iliamna Lake, length 120 km., Alaska Peninsula. The writer does not know whether they breed here.

REMARKS. Variability in skull, pelage color, and pelage pattern in the harbor seal is so great that one might, if one chose, describe an endless number of microgeographic races. The subspecies already described can hardly be separated on anatomical grounds; one must know first the source of the specimens. For example, after comparing *P. v. vitulina* with *P. v. concolor*, Doutt (1942, p. 114) concluded, "I am unable to state how specimens from the American side of the Atlantic can be distinguished from specimens taken on the European side."

Separation of Atlantic and Pacific specimens is perhaps a little easier: "In the Pacific coast skulls the premaxillae ascend not only to the nasals but extend posteriorly so as to touch the sides of the nasals for about 8 to 10 mm.; in the Atlantic coast specimens the premaxillae barely touch the nasals (in some cases do not quite reach them) . . . This distinction appears to be constant in all the skulls I have examined from the Alaskan and Kamschatkan coasts, as compared with those of the Atlantic coast"

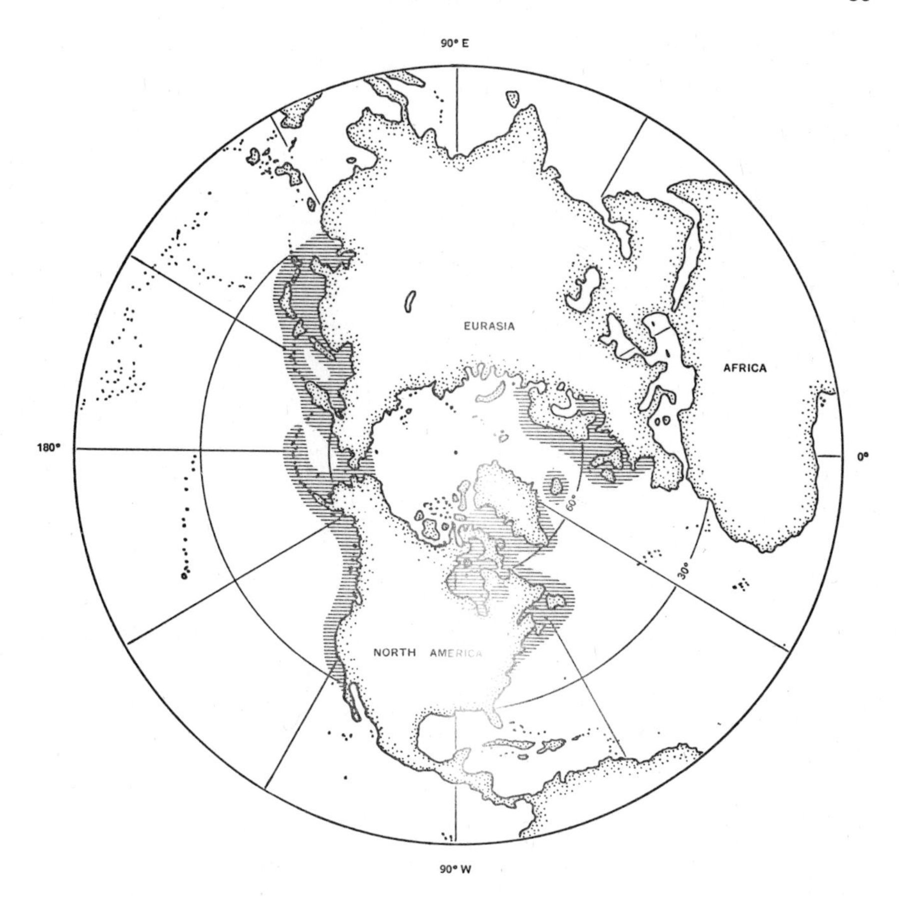

Fig. 6. Range of harbor seal. *Phoca vitulina vitulina,* eastern North Atlantic and Iceland; *Phoca vitulina concolor,* western North Atlantic; *Phoca vitulina mellonae,* Ungava Peninsula (Seal Lakes); *Phoca vitulina richardi,* eastern North Pacific and eastern Bering Sea; *Phoca vitulina largha,* western North Pacific, western Bering Sea, and Sea of Okhotsk.

(Allen, 1902*b,* p. 471). Doutt (p. 115) observed this character in 88 percent of specimens from the Pacific coast and in 14 percent from the Atlantic.

Within the North Pacific itself, Doutt stated (p. 115), "I am not able to present any characters by which specimens from the American and Asiatic sides . . . may be separated." Smirnov (1908, p. 63) applied the name *largha* to all harbor seals in the Pacific. Schwartz (1942, p. 222), on the basis of specimens in the U.S. National Museum, concluded that "the harbor seals from the Pacific coast of North America, south of the Alaska Peninsula, can be separated from *largha.*" But Osgood (1904, p. 48) had stated, "while admitting the probability that the seals of Bering Sea may differ subspecifically from those of Puget Sound, I am unable to appreciate any characters whatever after an examination of all the material now available."

All of the harbor seals which the present writer has seen on both coasts of North America and in European zoos were instantly recognizable as such. It seems best to regard them as a single species, *vitulina,* with five subspecies, as follows: *vitulina* (eastern Atlantic), *concolor* (western Atlantic), *richardi* (eastern Pacific), *largha* (western Pacific), and *mellonae* (Seal Lakes complex, Ungava Peninsula, Canada). The geographical barriers which separate the four subspecies are: (1) on the south, continents and warm-water regions; (2) on the north, ice-filled waters north of Siberia and Canada, rarely visited by harbor seals; and (3) wide expanses of open water in the northern North Atlantic, North Pacific, and Bering Sea. The assignment of a subspecific name to the population along each of the four arms of the distribution is tentative. Each name is a convenient label for a population which seems to have geographical, if not demonstrable morphological, distinction. For the fifth population the name *mellonae* is conserved, not because it stands for a remarkably different kind of seal but because it represents a situation: a population completely, though fairly recently, separated from all other harbor seals. The widest gap in the circumpolar distribution of the harbor seals is between eastern Novaya Zemlya and western Chukotsk Peninsula, a distance of about 2,500 km. (Naumov, 1933, p. 64).

Seton (1923, p. 86), Soper (1944, p. 237), and Harper (1956, pp. 74–75) have discussed local reports of fresh-water seals in various lakes and rivers of Canada. Certain of these seals are probably landlocked populations, others migratory populations, of *P. vitulina.*

Phoca vitulina vitulina Linnaeus, 1758. (See origin of name under species.)

RANGE. *General*: The eastern North Atlantic region (Bobrinskoi, 1944, p. 175; Ellerman and Morrison-Scott, 1951, p. 328). Novaya Zemlya (rare), Barents Sea, Murman coast (rare), apparently not in White Sea at all, Norway, Denmark, the Baltic Sea (but not, apparently, Gulf of Bothnia or Gulf of Finland), Netherlands, Germany, Belgium, France, Spain, and Portugal. Harbor seals have been recorded six times in Portugal, 41° N, during a century (Themido, 1947). While Pohle (1932, p. 78) stated that the harbor seal ranges as far north as Svalbard (Rossøya, 80° 50′ N), Saemundsson (1939, p. 6) stated that it is absent from "Spitsbergen and other islands of the Arctic Ocean." Pohle may have seen a ringed seal *Pusa hispida. Britain*: "The shallow waters and mudbanks of the Wash [East Anglia] are a sanctuary for thousands of common seals" (Darling, 1947, p. 9). About 400 were counted by Venables and Venables (1955) off Fitful Head, Shetland Islands, Scotland. Hentschel (1937, p. 45) stated that, of about 4,000 harbor seals on the Netherlands coast, 1,100 were killed annually. *Iceland*: common on all shores; Saemundsson estimated 12,000 harbor seals in 1939, of which about half were being taken annually. Denmark Strait is regarded as the midline of intergradation between *P. v. vitulina* on the east (Iceland) and *P. v. concolor* on the west (Greenland).

Phoca vitulina concolor DeKay, 1842.

Phoca concolor DeKay, 1842, p. 53. Long Island Sound, New York State.
Phoca (Phoca) vitulina concolor, Trouessart, 1904, p. 286.

TYPE. None. Species "based on New York examples of the light phase" (Allen, 1902*b*, p. 462). Locality given as Long Island Sound, near Sands Point, Nassau County, N.Y." by Miller and Kellogg (1955, p. 785).

RANGE. The western North Atlantic region (Dunbar, 1949). Eastern Greenland (fairly common up to Angmagssalik, occasional to Scoresby Sund, 70° N), western Greenland to Upernavik (fairly common in fjords of Godthaab and Frederikshaab), Ellesmere Island (northern limit 76° N? rare), southern Baffin Island, Southampton Island, Ungava Bay, Labrador. Anderson (1946) stated that the harbor seal is most common from Labrador to Maine, occasionally ascends the St. Lawrence River to Montreal; is recorded from Lake Ontario and the mouth of the Gatineau River near Ottawa. Brimley (1931) recorded stragglers from North Carolina (35° N).

Phoca vitulina mellonae Doutt, 1942.

Phoca vitulina mellonae Doutt, 1942, pl. 1 (frontis.) and p. 111. Lower Seal Lake, Ungava Peninsula.

TYPES. *Holotype*: Skin and skeleton, adult male, Carnegie Museum no. 15215, original no. 5112; collected in Lower Seal Lake (56° 30′ N, 74° 30′ W) on Ungava Peninsula, Quebec, on 23 March 1938, by J. Kenneth Doutt. *Paratype*: Skin and skeleton, skull broken, adult female, no. 15213. *Other specimens examined*: 2 skins-only (nos. 15211 and 15212), 1 embryo (no. 15214 out of the paratype), and 1 sealskin bag (no. 15216); all from the type locality (Doutt, *op. cit.* and *in lit.*).

RANGE. "Restricted to Upper and Lower Seal Lakes, which lie about ninety miles east of Richmond Gulf, Hudson Bay, Canada . . . the first race of *Phoca vitulina* to be described from an inland lake" (Doutt, *op. cit.* pp. 112, 114). Upper and Lower Seal Lakes communicate with each other; they are about 240 meters above sea level. Doutt has more recently described (1954, pp. 239–42) the environment and general habits of the Ungava fresh-water seals. He "would guess that the population . . . would be less than 1,000; probably 500 is nearer the maximum" (*in lit.*, 1957).

REMARKS. Doutt estimated that the subspecies *P. v. mellonae* was formed during a postglacial regime of 3,000 to 8,000 years' duration. It is distinguished from other races of *P. vitulina* (including *P. v. concolor* about 145 km. to the westward) by the slender coronoid process which reaches backward to the plane of the condyloid process; pelage darker than in 31 skins of *P. v. concolor* examined by Doutt, approaching the color of 6 skins of "*P. v. geronimensis*" (= *P. v. richardi*) from Baja California.

Phoca vitulina richardi (Gray) 1864.

Halicyon richardii Gray, 1864, p. 28. Vancouver Island, British Columbia.

Halichoerus antarcticus Peale, 1848, vol. 8, p. 30, pl. 5. Specimens actually from west coast of North America, mislabeled from Kerguelen Islands, Southern Ocean.

Phoca pealii Gill, 1866*a*, pp. 4, 13. Name substituted for *Halichoerus antarcticus*. California and Oregon (Territory).

Halicyon ? *californica* Gray, 1866*b*, p. 367. *Nomen nudum*. California.

Phoca (*Phoca*) *vitulina* (part), Trouessart, 1897, p. 385; *P. Richardsi* (*sic*) placed in synonymy.

Phoca richardii, Allen, 1902*b*, p. 491. Farther along in this paper Allen split *Phoca* into subgenera and split *richardii* into subspecies.

Phoca richardii pribilofensis Allen, 1902*b*, p. 495. St. Paul Island in Pribilof Islands, Bering Sea.

Phoca richardii geronimensis Allen, 1902*b*, p. 495. Isla San Jerónimo, 29° 47′ N, 115° 48′ W, Baja California, Mexico.

Phoca (*Phoca*) *vitulina richardsi*, Trouessart, 1904, p. 287.

Phoca vitulina richardii, Doutt, 1942, pp. 112, 120–21.

TYPES. *Syntypes*: (1) Skeleton, skull broken, of adult, British Museum (Natural History) no. 61.10.9.8, original no. 1431*a*; collected on Fraser River, British Columbia, in March (?) 1861, by surgeon Charles B. Wood; (2) skull only of subadult, no. 64.2.19.1, original no. 1431*b*; collected in Queen Charlotte Sound, British Columbia, in April or September (?) 1862, by Charles B. Wood (Scheffer and Slipp, 1944, pp. 374–75). Gray stated later (1874, p. 4) that the subadult skull had been the subject of his figures 1 and 4. For bibliographic purposes, the type locality is Vancouver Island, B.C.

RANGE. North and west coasts of North America from Herschel Island (69° 35′ N, 139° W, Dunbar, 1949, p. 9) to eastern Bering Sea, Aleutian Islands, and southward along the coast to northern Baja California, Mexico. Probably a rise in sea-water temperature near the latitude of Isla Cedros (28° 12′ N) marks the southern limit of the range. The writer saw no harbor seals in a survey of the shoreline of Isla de Guadalupe (29° N, 188° 16′ W) in June 1955. Although Guadalupe lies 260 km. off the mainland, one can be fairly sure that harbor seals have visited it from time to time. While representatives of three other genera (*Zalophus*, *Arctocephalus*, and

Mirounga) have colonized the island, why has not *Phoca*? Imler and Sarber (1947, p. 2) estimated "not less than 6,000 harbor seals living in the Copper River delta" of southern Alaska Peninsula. The Alaska Department of Fisheries (1956, p. 98) killed 21,853 seals here with depth bombs between 1952 and 1955.

REMARKS. In what region *P. v. richardi* intergrades with *P. v. largha* in the Bering Sea is not known. Schwartz (1942, p. 222) regarded the range of *P. v. largha* as including the Bering Sea and Arctic Ocean from at least as far north as Point Barrow to, and including, the Alaska Peninsula. The Pacific subspecies can generally be distinguished from the Atlantic "by the extension of the premaxillaries backward along the nasals" (Doutt, 1942, p. 113).

Phoca vitulina largha Pallas, 1811.

Phoca largha Pallas, 1811, p. 113. Eastern Kamchatka.

Phoca tigrina Lesson, 1827, p. 206. Kamchatka. Perhaps a harbor seal.

Phoca chorisi Lesson, 1828, p. 147. Bering Strait. Perhaps a harbor seal. Based on the "chien de mer du détroit de Behring" of Choris (1822, pl. 8) and said to range in Kamchatka, Aleutian Islands, and Kuril Islands. "Ses petits sont blancs comme la neige."

Phoca nummularis Temminck, 1847, p. 3. Japan. Allen (1902*b*, p. 466) regarded this as a ringed seal; Ellerman and Morrison-Scott (1951, p. 328) as a harbor seal.

Phoca (*Phoca*) *vitulina* (part), Trouessart, 1897, p. 384 (*largha* placed in synonymy); 1904, pp. 286–87 (*largha* disposed of with a question under four names).

Phoca ochotensis Allen, 1902*b*, p. 480. Mouth of Gizhiga River, Sea of Okhotsk. Not *P. ochotensis* Pallas, a ringed seal.

Phoca ochotensis macrodens Allen, 1902*b*, p. 483. Avachinskaya Guba, Kamchatka.

Phoca stejnegeri Allen, 1902*b*, p. 485. Bering Island, Commander Islands.

Phoca vitulina largha, Smirnov, 1908, pp. 2, 62.

Phoca vitulina largha natio *pallasii* Naumov and Smirnov, 1936, p. 177. Sea of Okhotsk.

Phoca petersi (part) Mohr, 1941, p. 58. Korea.

Phoca ochotensis kurilensis Inukai, 1942, p. 930. Southern Kuril Islands. Quite certainly a harbor seal (Scheffer, 1956).

TYPE. None. Species based on Russian stories about the "tschernaja nerpa" of eastern Kamchatka; locality "ad orientale littus Camtschatcae." The British Museum (Natural History) has topotypes from Petropavlovsk.

RANGE. From Bering Strait southwestward along Asiatic shores and islands to China; northwestward into Chukchi Sea (?). Rass *et al.* (1955, map on p. 109) showed *P. v. largha* distributed from East Cape (170° W) westward along the shores of Bering Sea, down the Kuril Islands, and completely around the shores of the Sea of Okhotsk. The breeding concentrations are in Sea of Okhotsk between latitudes 53° and 62° N. Barabash-Nikiforov (1938, p. 427) reported seals "near the Commander Islands the whole year. Parturition takes place at the end of April or beginning of May." N. G. Buxton (in Allen, 1903, pp. 162–63) described the hunting of harbor seals at many places in Sea of Okhotsk. Kuroda (1940, p. 23) stated that harbor seals arrive at the Seal Islands (Robben Island, off southeastern Sakhalin) about October, and are "not known south of these islands." However, Leroy (1940) reported that harbor seals go much farther south. He described and figured the skull of a young (yearling?) seal killed on Lung-hsü-tao k'ou, a small island at 37° 23′ N, 122° 40′ E, off Shantung Peninsula, on 25 March 1937. Furthermore, "fishermen told me that seals were coming every year at the beginning of spring but were unable to say how long they remain along the coast" (p. 67). He cited Fauvel (1892, p. 455 ff.) who stated, "Bien que je n'aie observé que le phoque commun (*Ph. vitulina*) le livres chinois décrivent quatre variétes de phocidés comme habitant les côtes du Nord . . ." The four kinds of phocids may be *Eumetopias, Zalophus, Callorhinus,* and *Phoca.* Harbor seals may actually range as far south as 32° N, for sea captains have told of seeing seals basking at the mouth of the Yangtze (Allen, 1938). Wilke (1954) reported the existence of a small industry on northern Hokkaido, largely dependent upon harbor seals. Two ships took 847 harbor seals in one month in early spring. The seals had been feeding mainly on *Theragra* and *Clupea.*

REMARKS. At least ten names have been coined—three in the past twenty years—for the harbor seal of the western North Pacific. Furthermore, the harbor seal has on several occasions been confused with the ringed seal *Pusa hispida,* somewhat similar in size and color pattern. Only the principal synonymy of *Phoca vitulina largha* has been given above.

P. v. largha is believed to differ from *P. v. richardi* in larger average size and in earlier pupping season (at least among northern populations). Wilke (*op. cit.*) stated that in Sea of Okhotsk the young are born from February to March, on pack ice, with the whitish, woolly embryonal pelage. This pelage is illustrated in Wilke's plate 1. On the contrary, there seem to be no records of *P. v. richardi* newborn young in the embryonal pelage (Scheffer and Slipp, 1944; Imler and Sarber, 1947). From as far north as the Pribilof Islands, near-term fetuses examined by the writer had already shed the white coat. (But a 36-pound pup taken on Pribilof pack ice in mid-April 1954—skin and skull now in the collection of K. W. Kenyon—is completely whitish and densely woolly. This specimen is thought to be a representative of *P. v. largha* carried by drifting ice south of its usual range.) Further information on distribution and characteristics of harbor seals along the edge of ice in Bering Sea and Okhotsk Sea is needed.

Genus **PUSA** Scopoli, 1777

Pusa Scopoli, 1777, p. 490. (ringed seal group)

TYPE. *Phoca foetida* Fabricius (= *Phoca hispida* Schreber). "The term [*Pusa*] was generically applied to what seems to have been *Phoca foetida*" (Allen, 1880, p. 414).

REMARKS. *Pusa* includes the ringed seal *P. hispida,* the Baikal seal *P. sibirica,* and the Caspian seal *P. caspica.* Of all the pinniped genera *Pusa* is perhaps the most difficult to fit into a taxonomic arrangement. It is a plastic group whose members have encircled the Arctic Ocean and have exploited both marine and fresh-water habitats. Wherever seals of the genus *Pusa* occur, however, they have retained a certain delicate structure of skull, and an affinity for ice, both of which set them apart from their nearest relative, the harbor seal *Phoca.*

The Baikal and Caspian seals have been regarded by certain authors as distinct species; by others as merely subspecies of *P. hispida.* The present writer lists them as species on the basis of their peculiar, relict position, their fish-eating (as against plankton-eating) habits, and their unspotted (or only lightly spotted) pelage. The landlocked members of the genus *Pusa* in northern Eurasia are, so far as known, confined to Lake Baikal, the Caspian Sea, and a cluster of small lakes near the Baltic. There are no pinnipeds (except rarely *Monachus*) in the temperate, brackish waters of the Black Sea and adjacent Sea of Azov; none in Aral Sea (ignore the map by Bartholomew *et al.,* 1911). If any pinnipeds (ringed seals?) were present in historic times in Lake Oron (= Ozero Oron), northeast of Lake Baikal in the Lena watershed, they no longer exist; while Nordquist (1899, p. 35) could find no records of seals in Lake Onega, northeast of Lake Ladoga.

Mohr (1952*b,* p. 160) has described intergeneric mating in captivity between a female ringed seal *Pusa hispida* and a male grey seal *Halichoerus grypus.* The single offspring was born dead. Subsequently, a female grey seal mated to a male ringed seal produced three successive young, all dead.

Pusa hispida (Schreber) 1775. (ringed seal)

Phoca hispida Schreber, 1775, Theil 3, pl. 86 (illustration and name); 1776, Theil 3, p. 312 (text with vernacular name). Greenland and Labrador.

Phoca foetida Fabricius, 1776 (Müller, p. viii). Greenland. *Nomen nudum.*

Phoca (*Pusa*) *foetida,* Trouessart, 1897, p. 386. *P. hispida* placed in synonymy.

Phoca (*Pusa*) *hispida,* Trouessart, 1904, p. 287. *P. foetida* placed in synonymy.

TYPE. None. Species based on the "neitsek" of the Greenland Eskimo, as described by the Danish missionary Cranz (Allen, 1880, pp. 616–17). Coasts of Greenland and Labrador.

RANGE (fig. 7). Circumboreal near the edge of ice, to the North Pole;

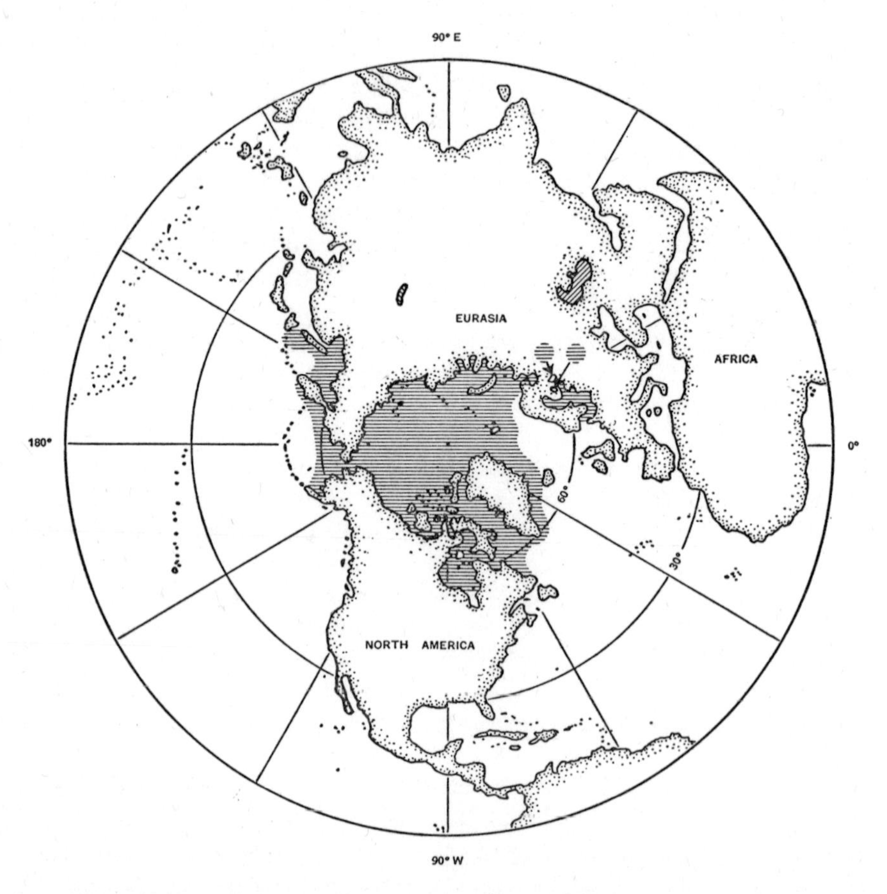

Fig. 7. Range of ringed seal group. (≡) Ringed seals: *Pusa hispida hispida,* north Atlantic, Hudson Bay, and Arctic Ocean; *Pusa hispida krascheninikovi,* Bering Sea; *Pusa hispida ochotensis,* Sea of Okhotsk; *Pusa hispida botnica,* Baltic Sea; *Pusa hispida ladogensis,* Lake Ladoga, U.S.S.R.; *Pusa hispida saimensis,* Lake Saimaa complex, Finland. (\\\\) Baikal seal *Pusa sibirica,* Lake Baikal. (///) Caspian seal *Pusa caspica,* Caspian Sea.

in landlocked lakes of extreme western Europe and in at least one lake on Baffin Island, Canada. South to France (one record) and Scotland (rare),

Hudson Bay and Labrador, Pribilof Islands (rare), Sea of Okhotsk, and northern Hokkaido.

Principal authorities: Naumov, 1933; Saemundsson, 1939; Doutt, 1942; Dunbar, 1949; Ellerman and Morrison-Scott, 1951; Mohr, 1952*b*; Miller and Kellogg, 1955; and Gijzen, 1956. Stefansson (1943, p. 702) mentioned the finding, by the Soviet explorer Papanin, of ringed seals and polar bears at the Pole in summer of 1937.

REMARKS. Many names have been coined for seals of the genus *Pusa*. At least one reviewer (Smirnov, 1929, p. 95) has employed a quadrinomial to identify a local population: *Phoca hispida pomororum* natio *rochmistrovi*, from the western coast of the White Sea. Another (Zukowsky, 1914, p. 230) has described a dwarf form *Pusa hispida pygmaea*, supposed to range from Novaya Zemlya to Svalbard in the same waters as the larger, more typical animal. Even after the Baikal and Caspian forms have been set aside as full species, twelve names for *Pusa hispida* remain in the collective lists of Ellerman and Morrison-Scott (1951), Mohr (1952*b*), and Miller and Kellogg (1955). (Other circumboreal pinnipeds, however, have been treated more conservatively. That is, of the bearded seal and walrus, two subspecies each are generally recognized.) In order to express the kinship of the sea-ice populations and at the same time to distinguish those in inland waters, the present writer would reduce the number of named forms of *Pusa hispida* to six. These include: one in the Arctic Ocean, three in peripheral seas (Okhotsk, Bering, and Baltic), and two in fresh-water lakes (Ladoga and Saimaa) barely, and fairly recently, cut off from the sea.

Pusa hispida hispida (Schreber) 1775. (See origin of name under species.)

Pusa hispida pygmaea Zukowsky, 1914, p. 230. Described from a living specimen 73 cm. in length in Carl Hagenbeck's Zoo, originally from Barents Sea, 77° 3′ N, 49° 40′ E.

Phoca hispida pomororum Smirnov, 1929*a*, p. 95. West coast of Novaya Zemlya.

Phoca hispida pomororum natio *rochmistrovi* Smirnov, 1929, p. 95. Sumski Posad, west coast of White Sea.

Phoca hispida birulai Smirnov, 1929, p. 96. Liakhov Island, New Siberian Islands.

Phoca hispida soperi Anderson, 1943, p. 27 (English); p. 39 (French). Near mouth of Takuirbing River (which drains Nettilling Lake), west coast of Baffin Island, Canada. See further discussion in Soper (1944, pp. 238–39).

Phoca hispida beaufortiana Anderson, 1943, p. 25 (English); p. 47 (French). Cockburn Point (68° 53′ N, 115° 03′ W), Dolphin and Union Strait, Canada.

RANGE. Arctic Ocean, occasionally to the Pole in open leads, and on sea ice along northern Eurasia, Greenland, and North America; southward to France (rare), Scotland (rare), Labrador and Hudson Bay or perhaps James Bay. According to Naumov (1933, p. 69) the ringed seal is the commonest and most widely distributed of all seals in the U.S.S.R. (The reason is clear; the U.S.S.R. has over 10,000 km. of ice-bordered coastline.) He stated further (pp. 75–76) that about 10,000 ringed seals are killed annually in northern Europe. Certainly the ringed seal is the most important of all pinnipeds in the economy of the natives of the far north. Saemundsson (1939, p. 8) reported that ringed seals do not breed on the shores of Iceland though they occasionally visit the northern shore (*i.e.*, Arctic Circle). Freuchen (1935) described at great length the habits of the "fjord seal" north of Greenland. He regarded it as the most common seal along the coasts of Labrador, Baffin Island, and Hudson Bay.

Pusa hispida ochotensis (Pallas) 1811.

Phoca ochotensis Pallas, 1811, p. 117. Near mouth of Gizhiga River in Sea of Okhotsk.

Phoca (Pusa) hispida gichigensis Allen, 1902*b*, p. 478. Near mouth of Gizhiga River.

Phoca (Pusa) hispida ochotensis, Ognev, 1935, p. 588.

TYPE. None. According to Allen (1880, p. 447), Pallas' *ochotensis* "presents a combination of characters thus far unknown in nature." However, the name clings to the ringed seal of the Far East. The type locality seems to be near the mouth of the Gizhiga River.

RANGE. Rass *et al.* (1955, map on p. 106) show the ringed seal as distributed around the shores of the Sea of Okhotsk, southward to the northern Kuril Islands and to the southern tip of Sakhalin; barely into Tatar Strait, west side of Sakhalin. N. G. Buxton (in Allen, 1903, p. 161) gave an account of ringed seal movements in the Sea of Okhotsk; as did also Naumov (1941, p. 73). Wilke (1954, p. 220) published a photograph of a ringed seal taken off northern Hokkaido in late April. According to Naumov (1933, pp. 75–76) about 50,000 ringed seals are killed annually in the Far East, including, presumably, the Bering Sea race.

REMARKS. Bobrinskoi (1944, p. 177) bracketed in the "Far Eastern group of small, light-colored" seals only the race *ochotensis*.

Pusa hispida krascheninikovi (Naumov and Smirnov), 1936.

Phoca hispida krascheninikovi Naumov and Smirnov, 1936, p. 182. Gulf of Anadyr, Bering Sea.

TYPE. Skin and skull of adult, sex unrecorded, Osteological Department, Zoological Institute, Academy of Sciences, Leningrad, no. 12,772, collected by Goudatti at Novo-Mariinsk Post, Gulf of Anadyr, Bering Sea, on 14 May 1897 (authors' English summary, 1936, p. 187, and K. K. Chapskiy, *in lit.*, 20 October 1957).

RANGE. Northern Bering Sea, intergrading in the vicinity of Bering Strait with *P. h. hispida*, and along the Kuril Islands with *P. h. ochotensis*; ranging southward with the pack ice to St. Michael, Alaska (63° 29′ N), and Bristol Bay. Around the Commander Islands solitary individuals appear in spring. During summer they are rarely seen, and in winter they are absent (Barabash-Nikiforov, 1938, p. 427). There seem to be no records for the Aleutian Islands, through which the present writer has traveled on four occasions. Ringed seals follow the ice as far south as northern Hokkaido where Japanese take a few commercially (Wilke, 1954).

Pusa hispida botnica (Gmelin) 1788.

Phoca vitulina [var.] *botnica* Gmelin, 1788, p. 63. Gulf of Bothnia, Baltic Sea.

Phoca annellata Nilsson, 1820, p. 365. Baltic Sea.

Phoca foetida var. *annellata*, Nordquist, 1899, p. 22.

Phoca hispida var. *annellata*, Smirnov, 1927, p. 20.

Phoca (*Pusa*) *hispida botnica*, Ognev, 1935, p. 582.

TYPE. None. Species based on the "Grå Sial" of Linnaeus' *Fauna Suecica*, 1747, according to Allen (1880, p. 616). Gulf of Bothnia, Baltic Sea.

RANGE. Baltic Sea, including Gulf of Bothnia and Gulf of Finland; occasionally in the Neva (length 30 km.) which connects the Gulf of Finland with Lake Ladoga (the habitat of *P. h. ladogensis*). The salinity of the Baltic falls very low at time of ice melt. Mohr (1952*b*, p. 188) stated that 8,968 ringed seals were presented for bounty in Finland in 1909.

REMARKS. Allen (1880, pp. 597–600) gave a long history of the synonymy of the ringed seal. He at first (p. 457) disposed of *botnica* as a synonym of "? *Phoca vitulina*"; later (p. 559) placed it under *Phoca foetida* Fabricius. Trouessart did not list the name *botnica*. Mohr (1952*b*, p. 187) has used *annellata* for the Baltic ringed seal while Bobrinskoi (1944, p. 177) has used *botnica*. Bobrinskoi included in the "Baltic group of medium-sized, dark-colored races" the seals of the Baltic Sea and gulfs, Lake Ladoga, and Saimaa complex.

Pusa hispida ladogensis (Nordquist) 1899.

Phoca foetida var. *ladogensis* Nordquist, 1899, p. 33. Lake Ladoga, Karelo-Finnish S.S.R.

Phoca (*Pusa*) *hispida ladogensis*, Trouessart, 1904, p. 288.

TYPES. *Syntypes*: In a table given by Nordquist are measurements of five skulls, nos. 6–9 and one unnumbered, from Sordavala (= Sortavala, 61° 42′ N, 30° 41′ E), "1820/VI 56 (Zoolog. Museum Helsingfors)." Paavo Voipio has kindly written (*in lit.*) that the following specimens of the Museum Zoologicum Universitatis, Helsinki, seem to represent four of the syntypes: Nos. 538 and 560, skin and skull, Sordavala, Lake Ladoga, 20.VI.1856, A. v. Nordmann, specimens missing. No. 3252, original no. 7, skull only, sex? collector? Lake Ladoga, 1885; figured by Nordquist in plate 1, fig. 1, and thus the principal type. No. 3253, skull only, Lake Ladoga, 1885, specimen missing.

RANGE. Lake Ladoga only, especially the deeper northern part. Seals occasionally swim up the Neva to Leningrad. Nordquist concluded that any mixing between the Ladoga and Baltic races must be unimportant since the Ladoga form is a very dark and distinct one. Grimm (1883, p. 45) stated that up to 1,000 seals were (in his time) killed annually in Lake Ladoga.

Pusa hispida saimensis (Nordquist) 1899.

Phoca foetida var. *saimensis* Nordquist, 1899, p. 28. Lake Saimaa, Finland.
Phoca (Pusa) hispida saimensis, Trouessart, 1904, p. 288.

TYPES. *Syntypes*: Nordquist stated (p. 29) that "the skull, hind feet, and skin, are the only parts of the Saimaa seal which I have seen." In his table are measurements of 9 skulls, original nos. 1–5 and 13–16; the longest with a CBL of 177 mm. There are, in the Museum Zoologicum Universitatis, Helsinki, 9 skulls and a skull fragment, all from Saimaa or Haukivesi, which appear to be the syntypes (Paavo Voipio, *in lit.*). The principal one is: skull only, no. 3255, original no. 5, figured by Nordquist in plate 1, fig. 1, Lake Saimaa, no sex, date, or collector's name.

RANGE. Principally Lake Saimaa; also a series of connected lakes: Haukivesi, Puruvesi, Orivesi, and Pyhäselkä; occasionally in smaller, isolated lakes in this vicinity. Saimaa lies at an altitude of 76 m. and is cut off from the sea. It drains into Lake Ladoga, 70 m. below it, by a stream too swift for seals to navigate. There are no valid records of seals in Lake Onega.

Pusa sibirica (Gmelin) 1788. (Baikal seal)

Phoca vitulina [var.] *sibirica* Gmelin, 1788, p. 64. Lake Baikal and Lake Oron (= Ozero Oron), U.S.S.R.

Phoca baicalensis Dybowski, 1873, p. 109, pls. 2 and 3 (also spelled *baikalensis*). Lake Baikal. The author did not include *sibirica* in his list of synonyms. Over fifty years later the author again published, though

Miller (1932, p. 150) concluded that there is ". . . no reason to cite any of these [names] he has subsequently [to 1926] used for seals."

Phoca (Pusa) sibirica, Allen, 1880, p. 464.

Phoca (Pusa) hispida sibirica, Trouessart, 1904, p. 288.

TYPE. None. Species based on the seal of Lake Baikal; habits described by Bell in 1763 and by Steller in 1774 (Allen, 1880, p. 612). For history of names, see Allen.

RANGE (fig. 7). Only in Lake Baikal; a fresh-water body about 650 km. long and 1,737 m. deep. It is 2,000 km. distant by air from Laptev Sea, quite surely the original source of the Baikal seal stock. According to Allen (1880, p. 612), Lake Oron (57° 20′ N, 116° 30′ E, northeast of Lake Baikal) was originally included in the range of *sibirica,* probably on the evidence of Steller. Russian zoologists agree that there are no seals in Lake Oron today.

Bobrinskoi (1944, p. 177) described the movements of seals in Lake Baikal as follows: "In winter the adult males are dispersed throughout the whole lake . . . immature individuals of both sexes usually on the west coast and adult females on the east. The seals (except pregnant females) spend the whole winter in the water, breathing at air holes that they make while the ice is thin and kept clear by constant use. Only pregnant females come out onto the ice . . . In the spring (starting in April) all the rest of the seals . . . come out onto the ice . . . forming aggregations of up to 100–200 individuals . . . At the beginning of summer, after the ice has finally broken up, the seals collect mainly in the northern part of the lake, to which drift ice is driven by the wind. At the end of June they start to congregate off the shores and to form summer rookeries on rocks projecting from the water. These finally break up with the formation of ice." Naumov (1933, p. 77) gave the annual kill at 5,000 to 10,000 seals. The total population is perhaps 40,000 to 100,000.

REMARKS. "Back a uniform olive brownish or brownish silver-grey colour, flanks and belly lighter and yellower; in rare cases spots. The young are yellowish white . . . Feeds on fish, chiefly gobies and 'golomyanka' (*Comephorus baicalensis*)" (Bobrinskoi, 1944, p. 177). The two important characteristics of *P. sibirica* are its almost immaculate pelt and its relict geographical position. It is here listed as a full species because it seems to be more clearly distinct from the sea-ice ringed seals than are *P. hispida ladogensis* and *P. hispida saimensis,* the other fresh-water ringed seals. A photograph of the skull was reproduced by Nordquist (1899, pl. 1, fig. 3). Chapskiy (1955*b*) has reviewed the history of the Baikal and Caspian seals but unfortunately the present writer has not seen an English translation of his work.

Pusa caspica (Gmelin) 1788. (Caspian seal)

Phoca vitulina [var.] *caspica* Gmelin, 1788, p. 64. Caspian Sea.

Phoca (*Pusa*) *caspica,* Allen, 1880, p. 464.

Phoca foetida subsp. *caspica,* Nordquist, 1899, p. 39, pl. 1, fig. 4; pl. 2, fig. 4; pl. 3, fig. 4.

Phoca (*Pusa*) *hispida caspica,* Trouessart, 1904, p. 288.

TYPE. None. Species "based on the account of the Caspian Seal given by Gmelin in 1770, in the third volume (p. 246) of his Reise durch Russland zur Untersuchung der Drey Naturreiche" (Allen, 1880, p. 426). For history of names see Allen, p. 609. The locality given by Gmelin (1788, p. 64) was "in mari, praesertim septentrionali, etiam Pacifico et Caspico."

RANGE (fig. 7). Only in the Caspian Sea. The Caspian is 1,280 km. long; has an area of 438,690 sq. km.; and its surface is 26 m. below sea level. The average salinity is 1.3 percent, or less than half that of sea water. The Caspian freezes in the northern part only, unlike Lake Baikal, which freezes entirely. Bobrinskoi (1944, p. 178) gave the following account: "During the summer the seals live in small herds chiefly in the central and southern parts of the Caspian. In the autumn a mass migration begins to the north and particularly the north-east of the Sea, where the seals form winter rookeries on the ice . . . The mass pupping period falls in January. The mating season is the end of February and first half of March . . . In May, most of the animals migrate southward . . . A very important commercial animal." Dr. Sergei W. Dorofeev told the writer in 1958 that the mean annual take during the twentieth century has been about 115,000 seals, and that the stock itself may include as many as 1,500,000 animals. Gmelin, perhaps in error, included Aral Sea in the range of this form, and was followed by Allen. However, Russian zoologists from the time of Grimm (1883) have agreed that there are no seals at all in Aral Sea.

REMARKS. Smirnov (1927, p. 17) gave a diagnosis of the species, including "colour in adult light, small-spotted, no ring shaped spots . . . Though it in other particulars diverges more or less from the other two [species *hispida* and *sibirica*], the skull remains typically *Pusa*-like." A photograph of the skull was reproduced by Nordquist (1899, pl. 1, fig. 4).

Genus **HISTRIOPHOCA** Gill, 1873

Histriophoca Gill, 1873, p. 179.

TYPE. *Phoca fasciata* Zimmermann.

REMARKS. The ribbon seal of the Pacific-Arctic, and its counterpart the harp seal of the Atlantic-Arctic, are the only pinnipeds with banded pelage. In this respect they stand apart from other seals which are uniform in color, or are broadly washed with several colors, or are spotted. In the ribbon and harp seals, too, the color pattern of the sexes is unlike, while it is like in two near relatives, the harbor and ringed seals. Smirnov (1927, p. 10) expressed the resemblance between ribbon and harp seals by listing them both under genus *Histriophoca.* Chapskiy (1955*a*, pp. 164, 188), no doubt with a similar idea in mind, proposed the subtribe Histriophocina. The breeding ranges of the ribbon and harp seals do not meet, being separated by archipelagos of the central Canadian-Arctic and Severnaya Zemlya.

Histriophoca fasciata (Zimmermann) 1783. (ribbon seal)

Phoca fasciata Zimmermann, 1783, vol. 3, p. 277. Kuril Islands, U.S.S.R.

Histriophoca fasciata, Pohle, 1932, p. 78.

TYPE. None. Species based on "Rubbon [*sic*] Seal Pennant I. p. 523" and said to inhabit the Kuril Islands.

RANGE (fig. 8). Fairly extensive—about 5,000 km. via water from end to end. Little is known about the movements of the ribbon seal. The center of abundance is northwestern Bering Sea. From Point Barrow (71° N, 171° W) southward, probably to winter limit of drift ice near the tip of Alaska Peninsula (55° N). Scammon (in Allen, 1880, p. 681) stated that ribbon seals occasionally visit Unalaska. Northwestward to eastern part of East Siberian Sea (Bobrinskoi, 1944, p. 175). Southwestward to Pribilof Islands (rare), Commander Islands (rare), all along the coast of Siberia from Bering Strait to the Kuril Islands and shores of the Sea of Okhotsk (Rass *et al.,* 1955, p. 111). The ribbon seal ranges on both sides of Sakhalin to its southern tip, and to Hokkaido. Wilke (1954) saw ribbon seals taken commercially near Hokkaido. Naumov (1941, p. 74) wrote that "this species is found in the Okhotsk Sea almost exclusively in the spring and early summer (April–June) on floating blocks of ice . . . In late summer . . . occasionally met with in the eastern part of the sea. In the spring these seals inhabit most frequently the eastern part of the Sakhalin shore, near the Tauisk bay [Tauyskaya Guba], and also in the northern part of the Sakhalin bay [Sakhalinskiy Zaliv] . . . most numerous in icy years. Seals with young are found only in the Tartar strait [Tatar Strait]. Young, still in their white embryo fur, were found on floating ice in March–May."

Genus **PAGOPHILUS** Gray, 1844

Pagophilus Gray, 1844, p. 3.

TYPE. *Phoca groenlandica* Erxleben.

REMARKS. Trouessart (1904, p. 287) proposed *Pagophoca* as a substitute for *Pagophilus,* assuming the latter to be a homonym of *Pagophila*

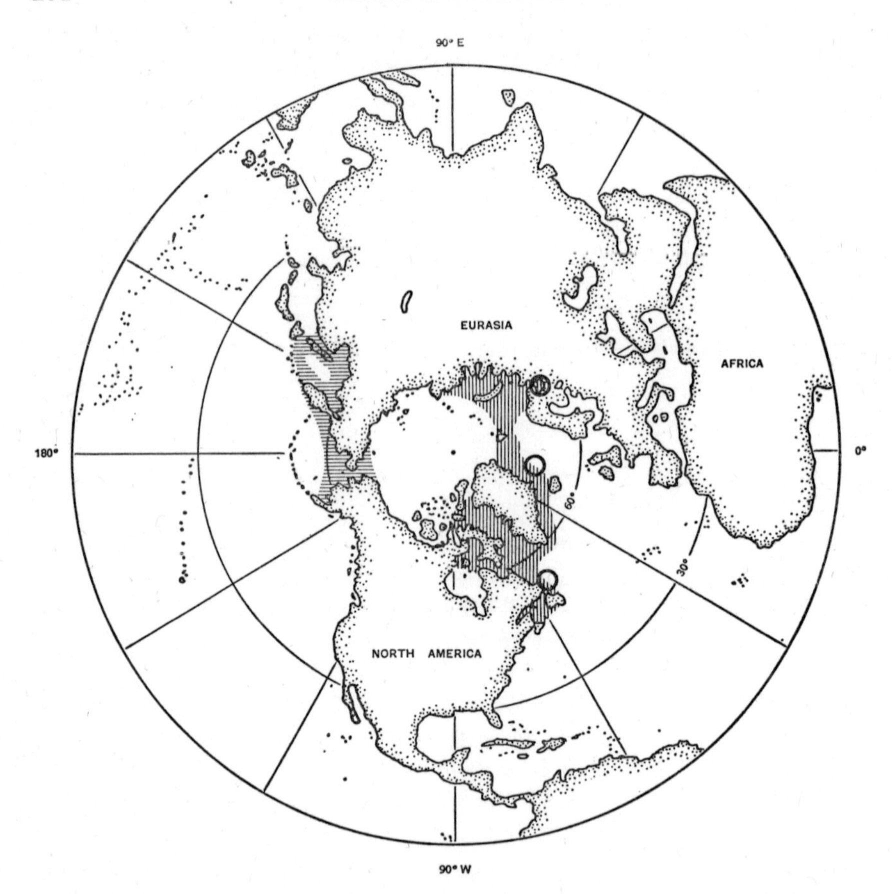

Fig. 8. Ranges of ribbon seal and harp seal. (≡) Ribbon seal *Histriophoca fasciata*. (|||) Harp seal *Pagophilus groenlandicus*. (O) Breeding sites of the three harp seal stocks: eastern (White Sea), central (Jan Mayen), and western (Newfoundland).

Kaup, 1829 (Aves). But Miller (1924, p. 165), Conisbee (1953, p. 64), and Miller and Kellogg (1955, p. 787) did not agree with Trouessart's interpretation. Naumov and Smirnov (1936, English summary on p. 187) preferred "*Philopagus*" to *Pagophilus* as grammatically nicer. *Pagophilus* is not yet universally accepted for the harp seal genus (*e.g.*, not by Chapskiy, 1955*a*, p. 161). According to J. A. Crook (*in lit.*) and Follett (1955, p. 10, par. 65*A*), the word *Pagophilus* is masculine.

From examination of a skull which seemed to combine certain characteristics of the harp seal *Pagophilus* and the ringed seal *Pusa*, Doutt (1942, p. 81) questioned, "Is it possible that the two species interbreed on rare occasions?" A seal killed on Corsica was supposed by Troitsky (1953) to be a cross between a Mediterranean monk seal *Monachus* and an arctic phocid, probably *Pagophilus*. But King (1956, p. 214) did not believe that the specimen differed in any way from typical *Monachus*, and the present writer agrees.

Pagophilus groenlandicus (Erxleben) 1777. (harp seal)

Phoca groenlandica Erxleben, 1777, p. 588. Greenland and Newfoundland.
Phoca groenlandica, Fabricius, 1776 (in Müller, p. viii, *nomen nudum*).
Pagophilus groenlandicus, Gray, 1850, p. 25, fig. 8.

TYPE. None. Based on several names—only that of Fabricius being Linnaean—and on travelers' accounts; locality "Groenlandia et Newfoundland."

RANGE (fig. 8). The harp seal is a deep-sea animal breeding on drifting pack ice in the North Atlantic and adjoining waters of the Arctic Ocean. It feeds on macroplankton and fishes. It is nowhere resident the whole year but makes long and regular migrations. The extreme range, in which limital records are rare, is from Severnaya Zemlya westward to the mouth of the Mackenzie River (almost to Alaska), through Kara Sea, White Sea, Norway, Svalbard, Jan Mayen, Iceland, Greenland, Labrador, Newfoundland, Hudson Bay, and Gulf of St. Lawrence; northward to latitude 75°–80° (Svalbard and southern Ellesmere Island); southward to about 50° N (Scotland, Germany, France), and 37° N (Cape Henry, Virginia). The last record was of a large individual, probably female with pup, held captive for a while on 12 March 1945. The range of the harp seal has been discussed by many authors, including Pohle (1932), Freund (1933), Saemundsson (1939), Bobrinskoi (1944), Dunbar (1949), Ellerman and Morrison-Scott (1951), Goodwin (1954), Miller and Kellogg (1955), and Fisher (1956).

The various breeding stocks and migration routes were summarized by Dunbar (pp. 14–16): "The harp seals breed on the ice in the early part of the year in three distinct regions: (1) off the northeast coast of Newfoundland and to the west in the Gulf of St. Lawrence; (2) in the Greenland sea between Iceland and Spitsbergen; and (3) in the White Sea. The summer distribution of these groups is probably as follows: (1) the west Greenland coast, Davis strait, Baffin bay and the Canadian eastern arctic; (2) east Greenland and the Greenland sea, and waters west of Spitsbergen; (3) east of Spitsbergen, and the waters round Novaya Zemlya . . . It is not known whether the three groups are genetically isolated the one from the other, or whether there is a greater or lesser degree of overlapping. On the basis of small differences in skull measurements, Smirnov [1927] has divided the species into three subspecies implying genetic isolation; but it is doubtful whether the numbers examined can give significance to the subdivision. There are also small differences in the season of parturtition which suggest divergent habits."

Smirnov (*op. cit.*, pp. 10–13) listed one species, *Histriophoca groenlandica,* and placed *oceanica* in synonymy. He then listed three "varieties," as follows:

"*H. groenlandica* [var.] *groenlandica*" (Fabricius) 1776, breeding in Labrador Stream off Newfoundland and in Gulf of St. Lawrence.

"*H. groenlandica* var.?" breeding in Jan Mayen Sea.

"*H. groenlandica* [var.] *oceanica*" (Lepechin) 1778, p. 259, pls. 6 and 7, breeding in White Sea and adjacent portions of Barents Sea.

Backer (1948, p. 61) mapped the three breeding grounds. In the opinion of the present writer, it is useful to identify the breeding stocks by locality names but not, on the basis of present information, by subspecific names. Further study will perhaps show that the three stocks of the harp seal resemble the three stocks of the northern fur seal (also highly gregarious and migratory) which exhibit no consistent anatomical differences.

Naumov (1933) stated that the average annual take of harp seals by Soviet and Norwegian sealers during the previous decade was 320,000. Through aerial photography he estimated over one million seals in the White Sea in 1927. The harp seal is at present the most important pinniped in the economy of the U.S.S.R. Fisher (1956) gave the average annual take of harp seals from the Newfoundland stock in 1949–54 as 229,000. He estimated that the total population of the Newfoundland stock (at the yearly peak) is 3,300,000. "[S. W.] Dorofejev has recently published in Rybnoje Khosjastvo an estimate of 5.5 million for all three stocks of harp seals . . . 3 million for the Canadian stock, 1.5 million for the White Sea stock and about 1 million for the East Greenland stock" (Johan T. Ruud, *in lit.*). Bertram (1940, p. 135) gave the annual world kill of harp seals as 500,000 a year, which agrees with the foregoing data. The world population at the summer maximum is perhaps 4.5–7 million.

Genus **HALICHOERUS** Nilsson, 1820

Halichoerus Nilsson, 1820, p. 376.

TYPE. *Halichoerus griseus* Nilsson (= *Phoca grypus* Fabricius).

REMARKS. Intergeneric mating in captivity between *Pusa* and *Halichoerus* was mentioned on p. 95.

Halichoerus grypus (Fabricius) 1791. (grey seal)

Phoca grypus Fabricius, 1791, p. 167, pl. 13, fig. 4. Greenland.

Halichoerus griseus Nilsson, 1820, p. 377.

Halichoerus grypus, Nilsson, 1841, p. 318.

TYPE. None. The "Krumsnudede Sael (*Phoca Grypus*)" of Fabricius was apparently based on an earlier account by Olafsen of the Greenland "Wetrar-Selur" or winter seal, so called because it gives birth in early winter (Allen, 1880, pp. 427, 431).

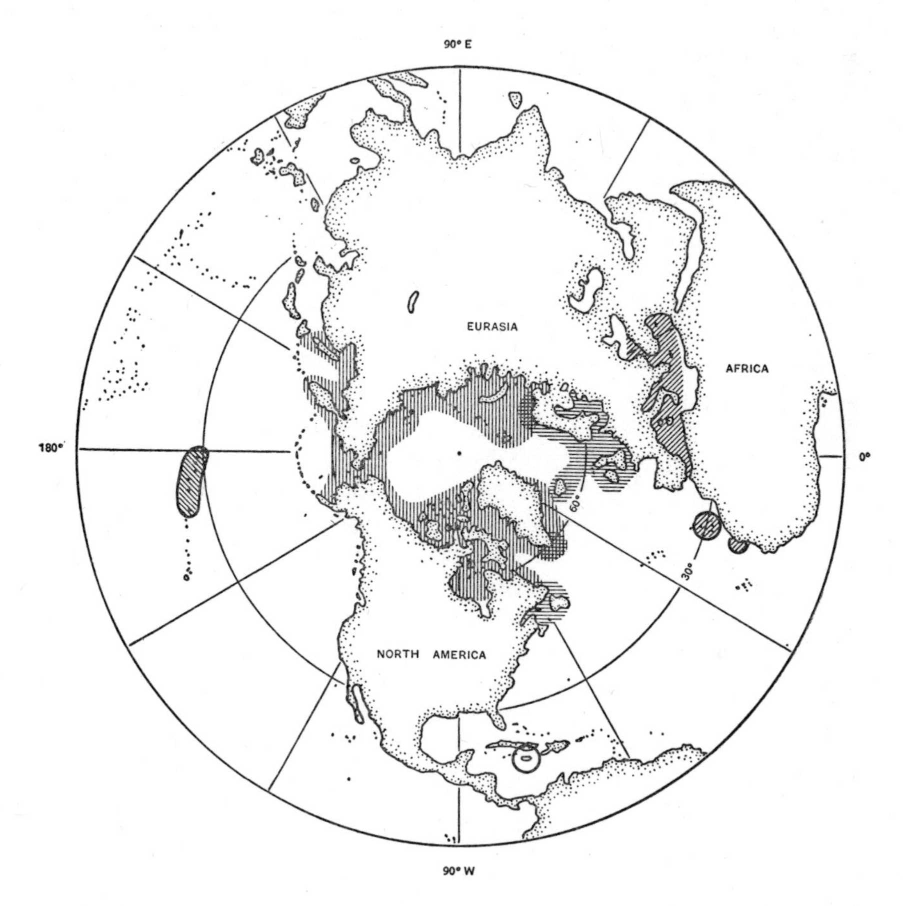

Fig. 9. Ranges of grey seal, bearded seal, and monk seals. (≡) Grey seal *Halichoerus grypus.* (|||) Bearded seal *Erignathus barbatus barbatus,* Atlantic-Arctic; *Erignathus barbatus nauticus,* Pacific-Arctic (boundaries of breeding range uncertain). (///) Mediterranean monk seal *Monachus monachus.* (\\\\\\) Hawaiian monk seal *Monachus schauinslandi.* (O) Caribbean monk seal *Monachus tropicalis* (exterminated ?).

RANGE (fig. 9). *General:* The center of abundance seems to be the British Isles. The grey seal ranges in temperate waters from Novaya Zemlya westward to Labrador, southward to France (rare) and New Jersey (40° S, rare). The most northerly breeding place in Europe may be the Murman coast (70° N, Bobrinskoi, 1944, p. 171). It breeds on the southwest coast of Svalbard (77° N) in the Gulf Stream influence; about 100 adults and 350 young are killed annually (Saemundsson, 1939). It is now rare in Greenland. The great depth to which grey seals may dive was mentioned on p. 21. *British Isles*: "The Atlantic grey seal is most common on the western and northern fringe of Scotland, more common than anywhere else in the world; for it is one of the rarest of seals. It has been my good fortune to have lived with them on North Rona, which is the breeding ground of perhaps half the world population of these seals" (Darling, 1947, p. 14). "In the Scillies and from the Lizard along the north coasts of Cornwall and Devon, on the Pembrokeshire coast, and then from the Clyde round Scotland to the Orkneys and Shetlands; less numerously down the east coast of Scotland and again at the Farne Islands; also on the Atlantic coasts of Ireland and, less numerously, on the east coast of Ireland . . . In general, Grey Seals prefer exposed, rocky coasts" (Morrison-Scott, 1952, p. 16). Commoner in Britain than the "common seal" *Phoca vitulina*. *Continental Europe*: "Novaya Zemlya [Pohle, 1932, p. 79], Barents Sea, Murman coast, neck of White Sea, Baltic Sea (including Finland, Gulf of Bothnia), Norway . . ." (Ellerman and Morrison-Scott, 1951, p. 332). Rarely seen in Netherlands (Gijzen, 1956, p. 21). No specimens in Portuguese museums (Themido, 1947), though da Gama (1957) described the grey seal as well as the harbor seal in her key to Portuguese mammals. *North America*: "Rarely and locally on Atlantic coast from Greenland to Labrador, Newfoundland, Nova Scotia, and Gulf of St. Lawrence where it is most common off Anticosti [Island] and Mingan Islands on North Shore; reaches southern limit on small islands near Rimouski, Riviere-du-Loup county, on south side of estuary of St. Lawrence River . . . Accidental off New Jersey" (Miller and Kellogg, 1955, p. 789).

Lockley (1954) has estimated the world population of grey seals over six months of age, as follows: Gulf of Bothnia (Sweden-Finland) 5,000; western Baltic–Kattegat (Denmark) 100; Norway 400; Faroes-Iceland 3,000; Gulf of St. Lawrence (Canada) 5,000; British Isles 20,000; total 33,500.

REMARKS. The writer visited the Farne Islands, eastern England, in April 1957 and concluded that many of the reefs are ideal situations for *Phoca vitulina*, yet Mrs. Grace Hickling (1956), an authority on the natural history of the islands, has seen only one harbor seal here in a number of years. Is the harbor seal unable to compete with the grey seal? A grey seal born and tagged in December on the Farne Islands was recovered less than two months later at Stavanger, Faeroe Islands, 885 km. away (Hickling, 1957, p. 111). Hewer (1957) has recently given an excellent description of social behavior in the grey seal.

Tribe ERIGNATHINI Chapskiy, 1955

Erignathini Chapskiy, 1955*a*, p. 164 (diagram), p. 165.

Genus **ERIGNATHUS** Gill, 1866

Erignathus Gill, 1866*a*, pp. 5, 9.

TYPE. *Phoca barbata* Fabricius (= *Phoca barbata* Erxleben).

Erignathus barbatus (Erxleben) 1777. (bearded seal)

Phoca barbata Erxleben, 1777, p. 590. North Atlantic Ocean.

Phoca barbata Fabricius, 1776 (in Müller, p. viii, *nomen nudum*). Greenland.

Erignathus barbatus, Gill, 1866*a*, p. 12.

TYPE. None. Species based on numerous accounts (Allen, 1880, pp. 662–66). A bearded seal was exhibited as early as 1743 in London. Miller (1924, p. 165) regarded the type locality as "coasts of Scotland, southern Greenland and Iceland."

RANGE (fig. 9). The bearded seal is circumboreal at the edge of ice, more or less resident, moving casually and not in regular migration; solitary or in groups of up to 50 on the ice in breeding season; preferring shallow waters 30–50 m. deep; occasionally swimming a few km. up rivers. Along all the coasts and islands of northern Eurasia and North America; northward to ice northwest of Novaya Zemlya (85° N, 80° E), southward to Norway (rare), British Isles (rare), France (rare), Newfoundland, eastern Bering Sea (Bristol Bay), northern Hokkaido, and Tatar Strait. The bearded seal is not uncommon on the north coast of Svalbard in winter but does not breed there. The bearded seal is of great importance to natives of the far north. Reported figures of the annual kill are certainly conservative: 2,000–4,000 (for Eurasia?) (Freund); 1,466 in 1931 in U.S.S.R. (Naumov).

Authorities: Pohle (1932, p. 79), Freund (1933, p. 77), Naumov (1933, p. 26), Saemundsson (1939), Bobrinskoi (1944, pp. 170–71), Dunbar (1949, p. 18), Ellerman and Morrison-Scott (1951, p. 331), Wilke (1954, p. 219), and Miller and Kellogg (1955, p. 788).

REMARKS. There are no compelling reasons for listing two subspecies, *E. b. barbatus* and *E. b. nauticus,* now firmly established in systematic literature. Bobrinskoi (1944, p. 171) regarded the two as "slightly differentiated geographic forms." In *nauticus* "the skull is the same length as in the typical subspecies but has a more massive, broad, and on the whole slightly shorter, facial section . . . The geographical boundary between the two forms apparently lies somewhere in Laptev Sea." (But where is its transpolar counterpart—the central Canadian–Arctic Archipelago?) The present writer feels, however, that it is useful to retain separate names for the Atlantic-Arctic and Pacific-Arctic groups. Similar treatment has been accorded another circumboreal, far northern pinniped: the walrus. In dealing with a circular distribution, whether marine or terrestrial, the taxonomist is often confronted with the necessity of weighing the importance of slight differences in anatomy and behavior of animals along a line which has no sharp beginning or end.

Erignathus barbatus barbatus (Erxleben) 1777. (See origin of name under species.)

RANGE. Eastern and western limits uncertain. Laptev Sea westward to central Canadian–Arctic Archipelago, including Poluostrov Taymyr, Poluostrov Yamal, Franz Josef Land, Barents Sea, White Sea, Svalbard, Iceland (visitor to north and west coasts), Greenland (except extreme north and east), Baffin Island, Hudson Strait and Ungava Bay (common), and Hudson Bay. Northward to ice northwest of Novaya Zemlya (85° N, 80° E). Southward (rarely) to Norway, British Isles, France, and Newfoundland (50° S). Freund (1933, p. 77) stated that at one time 2,000–4,000 bearded seals were killed annually in the Arctic Ocean.

Erignathus barbatus nauticus (Pallas) 1811.

Phoca nautica Pallas, 1811, p. 108. Sea of Okhotsk.

Phoca naurica (*sic*), Gray, 1871, p. 3. First distinction of a Pacific form as against *P. albigena* Pallas, 1811, p. 109, Atlantic.

Phoca (Erignathus) barbata (part), Trouessart, 1897, p. 383; 1904, p. 285. *Phoca "albigena* et *nautica,* Pall. Zoog., 1831" (read 1811) placed under *barbata.*

Erignathus barbatus nauticus, Osgood, 1904, p. 47.

TYPE. None. Species based mainly on the "lach-tak" of Steller (1751, p. 290), a Kamchadal name still in use. Naumov and Smirnov (1936, p. 185) gave the type locality as Penjina Bay, northern Sea of Okhotsk.

RANGE. Eastern and western limits uncertain. Central Canadian–Arctic Archipelago westward to Laptev Sea and Sea of Okhotsk, including Point Barrow, Chukotskiy Poluostrov, Commander Islands (rare), and Shantarskoye More. Southward to eastern Bering Sea (Bristol Bay), Tatar Strait (Zaliv Chikhacheva, 51° 25′ N), and northern Hokkaido (a few taken by sealers in early spring). Rass *et al.* (1955, p. 113) have published a map of the distribution of *E. b. nauticus* in western Bering Sea and Sea of Okhotsk. For other authorities, see under species.

"In the Sea of Okhotsk the bearded seal populates the entire sea, but is particularly abundant in shallow littoral zones. It prefers places full of creeks and bays. For the most part it leads a settled life, making but one or two local migrations during the year" (Naumov, 1941, p. 73). Temporary groups of 1,000 and more are formed. The seals give birth in February and mate in August (?).

REMARKS. While Smirnov (1927, p. 10) was earlier of the opinion that the bearded seal is "in general very constant . . . nowhere forming geographical varieties," he later (Naumov and Smirnov, 1936) concluded that in Pacific specimens "the skull is the same length . . . but has a more massive, broad, and on the whole slightly shorter, facial section." Hanna (1923) gave detailed measurements of a specimen 236 cm. in length, shot on one of the Pribilof Islands in January.

Subfamily MONACHINAE Trouessart, 1897

Monachinae Trouessart, 1897, p. 379. As presently regarded.
Stenorhyncina Gray, 1825, p. 340 (= *Monachus* plus *Hydrurga*).
Stenorhynchinae Gill, 1866*a*, pp. 6, 10 (= Lobodontini); Barret-Hamilton, 1902, p. 9 (= Monachinae).
Leptonyx Giebel, 1848, p. 290 (= *Lobodon, Leptonychotes, Ommatophoca,* and *Monachus*; or nearly Monachinae as now regarded).
Monachina Gray, 1869, p. 345 (= Monachini).
Ogmorhininae Turner, 1888, p. 63; 1912, p. 198 (= Monachinae).
Monachinae Kellogg, 1922, pp. 27, 87, and later authors, including Weber (1928, p. 352), Simpson (1945, p. 123), Frechkop (1955, p. 331), and King (1956, p. 251). In each case = Monachini.

REMARKS. This subfamily has no vernacular name. It includes the monk seals and the southern phocids (minus *Mirounga*). From the time they were placed together in 1825 by Gray, the two groups were alternately separated and rejoined by various writers over a period of nearly a century. In 1922 Kellogg proposed the subfamily Lobodoninae (as correct name for Stenorhynchinae Gill, 1866) and most writers have followed him in regarding the southern phocids as distinct. But Kellogg (pp. 88–89) agreed that the "Antarctic Lobodoninae . . . differ as much from each other as they do from *Monachus*" and "it is difficult to draw the exact differences that exist between the Monachinae and the Lobodoninae." In fact the only important difference between *Monachus* and the southern phocids, keeping in mind the great variation within the latter group, is one of geography. A separate subfamily for *Monachus* does not seem to be warranted. Characters shared by *Monachus* and the southern phocids are given in the key, p. 146, category 14.

The Monachinae (present sense) may have descended from upper Miocene *Monotherium* van Beneden (1876, pp. 800–801) of the North Atlantic and Mediterranean—a seal with six upper incisors (Kellogg, 1922, p. 87; 1936, p. 315). One branch remained in the subtropics and one pushed into the Southern Ocean, where its descendants now fill four distinct ecological niches.

Tribe MONACHINI. New tribe.

Monachina Gray, 1869, p. 345 (= Monachini). (monk seals)

REMARKS. Weber (1928, p. 352) listed under subfamily Monachinae the "eigentlichen Monachinae [= Monachini]" and "eine Abteilung . . . auch wohl als Unterfamilie Lobodoninae [= Lobodontini] abgetrennt." He thus suggested the present arrangement of the Monachinae into two tribes.

Genus **MONACHUS** Fleming, 1822

Monachus Fleming, 1822, p. 187, footnote. (monk seals)

TYPE. *Phoca monachus* Hermann. For history of names see King (1956, p. 210).

REMARKS. The genus *Monachus* contains three species: one in the Mediterranean Sea and adjacent waters, one in the Caribbean, and one in the Hawaiian Islands; northern limit in the Adriatic (45° N); southern in the Caribbean (13° N). Excepting the California sea lion which has been reported in latitude 21° 30′ N, monk seals are the only tropical pinnipeds. The total number of monk seals has remained low in historic times; it is now estimated (table 1) at 2,000–6,500. Almost nothing is known about the behavior of monk seals in the wild.

The three species of *Monachus* are geographically separated, one from another, by distances of at least 5,000 km. King (1956, pp. 226–34) has shown, curiously, that the Mediterranean and Caribbean species are anatomically more unlike than are the Caribbean and Hawaiian. This observation ties in with the hypothesis that monk seal progenitors—perhaps *Monotherium*—drifted westward from the Mediterranean to the Caribbean. "It is not difficult to accept the occupation of the West Indian islands from a source in the Mediterranean and along the Mauretanian coast, as the Canary Current passing down the latter coast would bear the emigrants to the eastern limit of the North Equatorial Current sweeping due west to the Caribbean Sea. The extension of the range to Hawaii and across the Isthmus of Panama is feasible when it is accepted that Phocids are capable of considerable overland journeys . . . The Isthmus of Panama at its narrowest is much less than fifty miles, and its lowest height above sea level less than 200 feet. If rivers were exploited by the seals the distances travelled overland might have been still further diminished. The North Equatorial Current could well have borne the animals to the islands they now occupy" (King, 1956, pp. 222–23). On the other hand, Kellogg (1922, p. 88) suggested that "the Monachinae must have had a very extensive distribution in the tropical seas as early as or even before the Lower

Miocene [when there was a waterway connecting the Caribbean and the Pacific]."

Differences among *M. monachus, M. tropicalis,* and *M. schauinslandi* are slight, and the three forms would be regarded as subspecies if it were not for the evidence of their complete isolation. Smirnov (1927, pp. 6–7) and King (1955, pp. 226 ff.) gave characters for separating *M. monachus* from the other two species. Paraphrased and arranged in the form of a key, these are:

1. Lacrymal bone well developed, forming an anteorbital process; skull broader (MW about 1/1.76 CBL); fronto-maxillary suture longer than naso-maxillary; incisors without pronounced "waist" at junction of root and crown; molars set obliquely; molars with large central cusp and single smaller anterior and posterior cusps; incisor row slightly curved; pelage often with ventral white patch.................*M. monachus.*

2. Lacrymal bone does not form an anteorbital process; skull narrower (MW about 1/2 CBL); fronto-maxillary suture shorter than naso-maxillary; incisors with pronounced "waist"; molars set in line; molars with low central cusp, a single smaller anterior and two small posterior cusps; incisor row straight; pelage always (?) without ventral white patch...........................*M. tropicalis* and *M. schauinslandi.*

Monachus monachus (Hermann) 1779. (Mediterranean monk seal)

Phoca monachus Hermann, 1779, p. 501, pl. 12 (entire animal), pl. 13 (flippers and face). Osor, Adriatic Sea.

Phoca albiventer Boddaert, 1785, p. 170. Osor, Adriatic Sea. Description based on the captive seen by Hermann.

Monachus monachus, Turner, 1888, p. 67.

TYPE. *Holotype*: A male captured at Osor on Cres Island, northern Adriatic Sea, Yugoslavia, in 1777; exhibited alive in Paris and Strasbourg, France; specimen not saved.

RANGE (fig. 9). Monk seals are thinly scattered along the Anatolian coast of the Black Sea (42° N) and Adriatic Sea (45° N); coasts and islands of the Mediterranean (except most of the Mediterranean coast of North Africa); southward to Cap Blanc, Spanish West Africa (21° N), with Canary Islands and other offshore islands (King, 1956, maps, figs. 1 and 2; Bobrinskoi, 1944, p. 170).

Monachus tropicalis (Gray) 1850. (Caribbean monk seal)

Phoca tropicalis Gray, 1850, p. 28. Pedro Cays, 80 km. south of Jamaica.

Monachus tropicalis Gray, 1866*b*, p. 20.

Cystophora antillarum Gray, 1849, p. 93. Skin and skull of a "very young seal" from "West Indies," poorly described, no figures. Quite certainly

= *C. cristata.* Allen (1880, pp. 719–20) believed that Gray had mistaken the locality; though Miller (1917) later reported an immature hooded seal killed not far from here (Cape Canaveral, Florida, 28° 27′ N, 80° 32′ W).

TYPE. *Holotype*: An imperfect skin, later stuffed, without skull, British Museum (Natural History) no. 1847.2.2.2; collected in 1846 at Pedro Cays, 80 km. south of Jamaica, by George Wilkie, who gave it to P. H. Gosse, who gave it to the museum (King, 1956, pp. 215–16; Gray, 1849, p. 93; 1866*b*, p. 20). According to Gosse (in Allen, 1880, p. 715) the type animal measured 6.5 ft. (198 cm.) in length from nose to tip of tail (subadult).

RANGE (fig. 9). Formerly shores and islands of the Caribbean Sea and Gulf of Mexico, including Central America but probably not South America; now nearly extinct; two individuals seen in 1949 near Jamaica (King, 1956, map fig. 3); known to occur in Jamaican waters in 1952 (Palmer, 1954, p. 162). Northern and southern limits along Texas (28° N) and Honduras (13° N). The downfall of the Caribbean monk seal started during the second voyage of Columbus, when a ship's crew killed seals at Alto Velo (17° 29′ N, 71° 38′ W), just south of Haiti, in August 1494.

Monachus schauinslandi Matschie, 1905. (Hawaiian monk seal)

Monachus schauinslandi Matschie, 1905, p. 258. Laysan Island, Hawaiian Islands.

TYPES. The types were quite certainly collected in summer of 1896 when Dr. H. H. Schauinsland, wife, and biological assistants spent June–September on Laysan. In one account (1899, p. 64) he stated simply that "Robben kommen vereinzelt, wenn auch recht selten, an der Insel vor." The specimens are not labeled as to date or sex.

Holotype: A skull and skin, stuffed, Zoologisches Museum Humboldt-Universität zu Berlin no. 32795; collected by Schauinsland on Laysan Island, Hawaiian Islands, in summer of 1896 (?). From the reported length of the skull, CBL 265 mm., and from King's table (1956, p. 246) it is supposed that the holotype is an adult but not full-grown animal. *Paratypes*: S. H. W. Stein wrote (*in lit.*) that with this skull there are three other specimens of the same species in the Schauinsland collection, none designated as paratypes, though mentioned in the original description. These are: no. 32796 "Schädelteil," no. 32797 "Schädelteil," and no. 32798 "Kopf- und Halsfell."

RANGE (fig. 9). Hawaiian Islands between 20° and 30° N; mainly the northwestern islands including Kure Island (= Ocean Island), Midway Islands, Pearl and Hermes Reef, Lisianski Island, Laysan Island, and French Frigate Shoals; an extent of about 1,600 km.; straggling to Hawaii

(King, 1956, map fig. 4). While the entire population numbers only about 1,000 to 1,500 (K. W. Kenyon, *in lit.*), it has been since 1909 under nominal protection of the United States and is in no immediate danger (Kenyon and Scheffer, 1955, p. 27). The population may once have been larger. The ship *Gambia* reportedly brought 1,500 skins to Honolulu in 1859 (Atkinson and Bryan, 1913). Kenyon, however, has studied the source of this report and has concluded that little confidence can be placed in it.

Tribe LOBODONTINI. New tribe.

Stenorhyncina Gray, 1825, p. 340 (= *Monachus* plus *Hydrurga*).
Stenorhynchinae Gill, 1866*a*, pp. 6, 10 (= Lobodontini).
Lobodontina Gray, 1869, p. 345 (= *Lobodon*).
Lobodoninae Kellogg, 1922, pp. 27, 89; Weber, 1928, p. 352 (= Lobodontini).
Lobodontinae Hay, 1930, p. 562; Simpson, 1945, p. 122; Frechkop, 1955, p. 328; King, 1956, p. 251. In each case = Lobodontini. See history of names under Monachinae and Monachini, pp. 111 and 112.

REMARKS. This group has no vernacular name, though "antarctic phocids" is often used with the understanding that the elephant seal *Mirounga leonina* is excluded. The Lobodontini include four monotypic genera: *Lobodon* (crabeater seal), *Ommatophoca* (Ross seal), *Hydrurga* (leopard seal), and *Leptonychotes* (Weddell seal). These are distributed in the Southern Ocean from the northern edge of pack ice (August and September; about 55° S on the Atlantic side, 60° S on the Indian Ocean side, and 65° S on the Pacific side) southward to the Bay of Whales in Ross Sea (80° S). One form (*Leptonychotes*) remains near fast ice along the Antarctic Continent, others (*Lobodon* and *Hydrurga*) feed and breed along the outer edge of pack ice and wander as far north as temperate waters of the major continents (South America and Australia, about 35° S). *Ommatophoca* remains, so far as known, near the outer edge of the pack ice.

Mackintosh and Herdman (1940) and Deutsches Hydrographisches Institut (1950) have shown by map the distribution of pack ice in the Southern Ocean. *Antarctic Pilot* (1948) gave briefly the status of seals on certain high-latitude shores and islands. Roberts (1948) listed about 200 antarctic expeditions and excluded many others of a purely commercial nature (sealing and whaling).

Certain patterns of distribution are known for the individual species of Lobodontini. Where a species appears to be rare or absent, the explanation may stem either from poor information about the region or from actual scarcity of seals because of a bleak environment. The explanation will in no case be overkilling by man, since the Lobodontini are as yet virtually unexploited. New, important concentrations of seals will probably not be discovered in the antarctic, for its waters have by now been fairly well crisscrossed.

A clue to the antiquity of the Lobodontini has been suggested by Markowski (1952, p. 149): "It seems from the literature that the Pseudophyllidean Cestodes found in the Antarctic seals do not occur in any other species of Pinnipeds."

Genus **LOBODON** Gray, 1844

Lobodon Gray, 1844, p. 2.

TYPE. *Phoca carcinophaga* Hombron and Jacquinot.

REMARKS. Gray (*op. cit.*), Allen (1905, p. 92), and certain other writers have used the feminine form "*carcinophaga*" with this generic name. "*Lobodon,*" however, is a combination of two nouns, one masculine and one neuter. It seems wise to follow the example of Bertram (1940, p. 84) and regard "*Lobodon*" as masculine.

Lobodon carcinophagus (Hombron and Jacquinot) 1842. (crabeater seal)

Phoca carcinophaga J.-B. Hombron and H. Jacquinot, 1842 (in C. H. Jacquinot, pls. 10 and 10*a*), Scotia Sea about 60° S, 35° W, Southern Ocean. The colored plates of Hombron and Jacquinot are captioned "*Phoca carcinophaga*"; the atlas in which they appear is dated 1842–53; the plates are thought to date from 1842. The authors of the atlas are not named on the title page. In 1853, a description of the crabeater was published in the same series (Zool., T. III, Mam. et Ois., p. 27) by H. Jacquinot and J. Pucheran under the name "*Lobodon carcinophaga,*" for the reason that Gray had, in 1844 in the interval between the publication of the plates and the description, established the genus *Lobodon*. Jacquinot and Pucheran (*op. cit.*) credited authorship of the plates to Hombron and Jacquinot, and since J.'s and P.'s statement was almost contemporary, it should be taken as authentic. In summary, the first use of the name *carcinophaga* coupled with a description (colored plates of animal and skull) was by J.-B. Hombron and H. Jacquinot, about 1842.

Lobodon carcinophaga, Gray, 1844, p. 2.

Ogmorhinus (Lobodon) carcinophagus, Trouessart, 1897, p. 381; 1898, p. 1303; 1904, p. 284. Credited to Hombron and Jacquinot, 1842.

Lobodon carcinophagus, Berg, 1898, p. 15.

TYPE. *Holotype*: A skin and skull collected between the South Sandwich Islands and the South Orkney Islands (= Powell Islands) at a distance of 150 leagues from either, during the 1837–40 expedition of the corvettes *L'Astrolabe* and *La Zélée* to the Southern Ocean; figured in plates 10 and 10*a* of Hombron and Jacquinot; not described. Specimen still in existence?

Barrett-Hamilton (1902, p. 36) stated that a skull, now Cambridge University Museum of Zoology no. 897, was purchased in 1853 from a member of the *Astrolabe* party. This skull, however, was from the expedition of 1826–29, not 1837–40 (both under Dumont d'Urville).

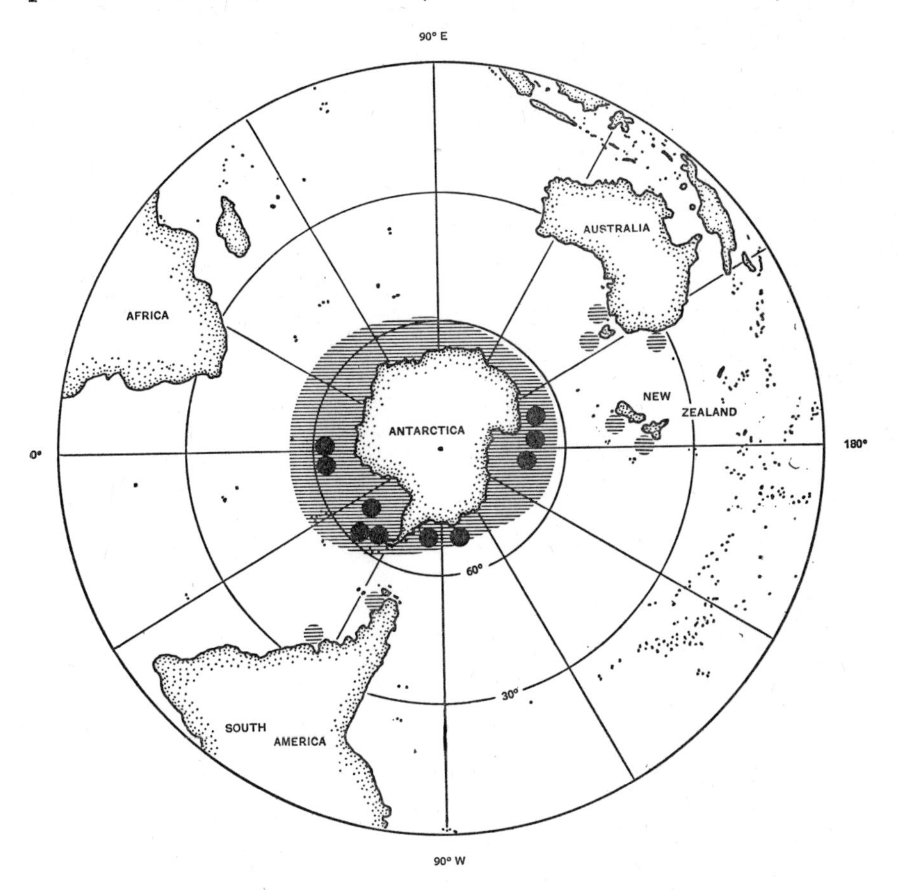

Fig. 10. Ranges of crabeater seal and Ross seal. (≡) Crabeater seal *Lobodon carcinophagus*, with limital records from continents. (●) Ross seal *Ommatophoca rossi*, all known records (each record may include several seals and several nearby sites).

RANGE (fig. 10). *General* (largely from Bertram, 1940, pp. 85–91): The crabeater is gregarious, semimigratory or perhaps migratory, pelagic, with least attachment of any of the Lobodontini to the shore. It early received the name "crabeater" whereas "krilleater" would have been more appropriate. Circumpolar in the Southern Ocean; following the edge of pack ice. While a few seals may winter beneath sea ice, most of them move outward from Antarctica in autumn to, roughly, latitudes 55°–65° S, where they spend spring and summer. Pronounced local movements of seals have been observed, and Bertram (p. 87) stated, ". . . it is probable that, when there is a greater knowledge of the animal spread throughout the year, it will be necessary to class the species with the truly migratory seals." Movements of seals of reproductive age are perhaps distinct from those of younger individuals. Large numbers of seals may appear suddenly in spring. According to Bertram (p. 85), "the Crabeater seal is by far the most abundant of the four species of truly antarctic seals . . . It is . . . at least semi-gregarious in its habits . . . In view of its wide, circumpolar range, it is probable that the Crabeater is the most abundant of all the Pinnipedia." Crew of the *Balaena* killed 4,000 to 5,000 off Louis Philippe Land in the 'nineties (Rudmose Brown, 1913, p. 194). Marr (1956) has shown in an instructive paper that whale and crabeater food is not at all uniformly distributed in the Southern Ocean.

Limital range: On the continent of Antarctica "thirty miles from the sea shore and 3,000 feet above sea level, their carcasses were found on quite a number of occasions" (Wilson, 1907, p. 34). *New Zealand*: 1 or 2 seen each winter (Bertram, 1940, p. 85). St. Kilda, Melbourne, July 1897 (Hall, 1903). Wanganui Heads, 39° 56′ S, and Petone Beach, 41° 14′ S (Oliver, 1921). *Australia*: New South Wales, 1 alive on Manly Beach in early July 1929 (LeSoeuf, 1929). Victoria, Portland, January 1894 (Hall, 1903). *Tasmania*: First record, a carcass at Ralph Bay, 7 September 1945 (Anonymous, 1946*b*). *South America* (Berg, 1898; Vaz Ferreira, 1956*a*): The South American records include 3 from near the mouth of the Río de la Plata, Uruguay-Argentina, 35° S, and one from Río Santa Cruz, southern Argentina, 50° S. *South Africa*: No records.

Genus **OMMATOPHOCA** Gray, 1844

Ommatophoca Gray, 1844, p. 3.

TYPE. *Ommatophoca rossii* Gray.

REMARKS. Barrett-Hamilton (1902, p. 51) concluded that, while the skull of *Ommatophoca* cannot be confused with that of any other recent pinniped, it somewhat resembles the skulls of other antarctic phocids as well as the skull of *Cystophora*. "Altogether, I can only regard *Ommatophoca* as a most interesting generalised form, an annectant genus, showing affinity both to the Stenorhynchinae [= Lobodontini] and the Cystophorinae, to both of which it stands in a quasi-ancestral relationship. I think it more convenient that it should remain included in the Stenorhynchinae than that it should form the type of a new family or subfamily." And (on p. 11) "if any convenient result would accrue therefrom, it might be taken to constitute a separate sub-family the *Ommatophocinae,* but I do not see how the multiplication of sub-families can help us much."

Ommatophoca rossi Gray, 1844. (Ross seal)

Ommatophoca rossii Gray, 1844, p. 3 (*nomen nudum*), pp. 7–8, pls. 7–8. Ross Sea, Southern Ocean.

TYPES. Collected in the Ross Sea during the British Antarctic Expedition of 1839–43 in the ships *Erebus* and *Terror* under James Clark Ross. Information on the types is from labels on specimens, from Gray's original account, from his hand list (1874), and from Barrett-Hamilton's (1902) study of the original 2 specimens plus 6 from the much later *Southern Cross* expedition. Barrett-Hamilton (pp. 46, 66) estimated the age and sex of the two syntypes, designated one as the "type" (= lectotype), and perhaps added to the label of this one a locality notation (pack ice, north of Ross Sea, 68° S, 176° E). *Lectotype*: Skeleton (of adult female?), British Museum (Natural History) no. 1843.11.25.4, original no. 324*a,* figured by Gray in pl. 7 (entire animal in life pose) and pl. 8, fig. 1 (skull), fig. 2 (cranium), and fig. 4 (5 teeth); also in 1874, fig. 9 and pl. 11. Skull damaged. The skin of this animal is apparently lost. *Syntype*: Skull only (of immature female?), no. 1843.11.16.7 (last digit given as "1" by Gray, 1874), original no. 324*b,* figured by Gray (1844) in pl. 8, fig. 3 (palatal region), and fig. 5 (4 teeth). Skull fragmentary.

RANGE (fig. 10). The Ross seal feeds along the edge of pack ice in the Southern Ocean; probably does not winter under ice. There are no records from island or continental beaches and none from the vast semicircular edge of Antarctica, including about 155 degrees of angle, which fronts the Indian Ocean. Records include (east to west from the meridian of Greenwich): off Queen Maude Land (individuals seen between 5° E and 9° W, according to Laws, 1953*a*); Falkland Islands Dependencies (between South Orkney Islands and Coats Land, 40° W; Scotia Bay, 45° W; Joinville Island and Louis Philippe Peninsula, 56° W; Deception Island, 60° W); Bellingshausen Sea (a *Belgica* record cited by Allen (1942, p. 458) as "83–85° E" should read "W"); off Thurston Peninsula (101° W); Ross Sea and vicinity (Discovery Inlet, 171° W; type locality, 176° E?; Drygalski Ice Tongue, 164° E; perhaps increasing in numbers toward Balleny Islands, 163° E; off Victoria Land, 160° E. Sapin-Jaloustre (1953, p. 12) reported no Ross seals seen in two expeditions to Adélie Coast (140° E).

Bertram 1940 (opp. p. 134) wrote of the Ross seal that "less than 50 have ever been seen." Laws (1953*a*) estimated the total number in Falkland Islands and Falkland Islands Dependencies at 10,000. Concentrations of Ross seals for breeding purposes have not been seen, although Wilson (1907, p. 42) stated that young animals had been taken in the South Orkneys by the Argentine Expedition of 1903.

Genus **HYDRURGA** Gistel, 1848

Hydrurga Gistel, 1848, p. xi.

Stenorhinchus É. Geoffroy Saint-Hilaire and F. Cuvier, 1826, p. 549. Preoccupied in Crustacea and Insecta. For history of the name *Hydrurga* see Allen (1905, p. 86).

TYPE. *Phoca leptonyx* Blainville.

Hydrurga leptonyx (Blainville) 1820. (leopard seal)

Phoca leptonyx Blainville, 1820, p. 298. Vicinity of Falkland Islands, South Atlantic Ocean.

Ogmorhinus (Ogmorhinus) leptonyx, Trouessart, 1897, p. 380; 1904, p. 283.

Hydrurga leptonyx, R. I. P[ocock], 1902 (in Barrett-Hamilton, p. 26).

TYPES. History of the types has been assembled from Blainville (*op. cit.*), Flower (1884, p. 212), Barrett-Hamilton (1902, pp. 25–27), Allen (1905, pp. 86–87), Hamilton (1939*a,* p. 243), and Patrice Paulian (*in. lit.*). *Syntypes*: (1) Young male, stuffed skin (7–8 ft. long) destroyed in 1914, and skull, in the collection of M. Hauville at Le Havre; later transferred to Paris; now Muséum National d'Histoire Naturelle, Laboratoire d'Anatomie Comparée, no. A 3578; damaged in basioccipital region; from vicinity of Falkland Islands; described by Blainville (pp. 297–98) and figured (fig. 5, skull). This skull is the principal type.

(2) Skull, Museum of the Royal College of Surgeons of England, Osteological Collection no. 3938 (no. 1091 in Flower's catalogue); brought by William Keane in a whaler from South Georgia; described but not figured by Blainville (pp. 287–88); not in British Museum in 1957; supposedly lost in World War II.

RANGE (fig. 11). The leopard seal is perhaps the most widely distributed of the Lobodontini; solitary, nowhere abundant; recorded from scattered points around Antarctica; occasionally from southern tips of the other southern continents. The most northerly record is from Lord Howe Island (31° 31′ S). Leopard seals migrate northward to ice-free islands in winter. The ability to subsist on many kinds of food—penguins, fishes, squids, seals, and carrion—permits the leopard seal to range widely. Principal records (east to west from the meridian of Greenwich):

Falkland Islands and Falkland Islands Dependencies: Laws (1953*a*) estimated about 40,000 leopard seals, of which fewer than 80 a year are killed by man. Specific records include South Sandwich Islands, where the *Norvegica* collected specimens in 1928–29 (Sivertsen, 1954, p. 54);

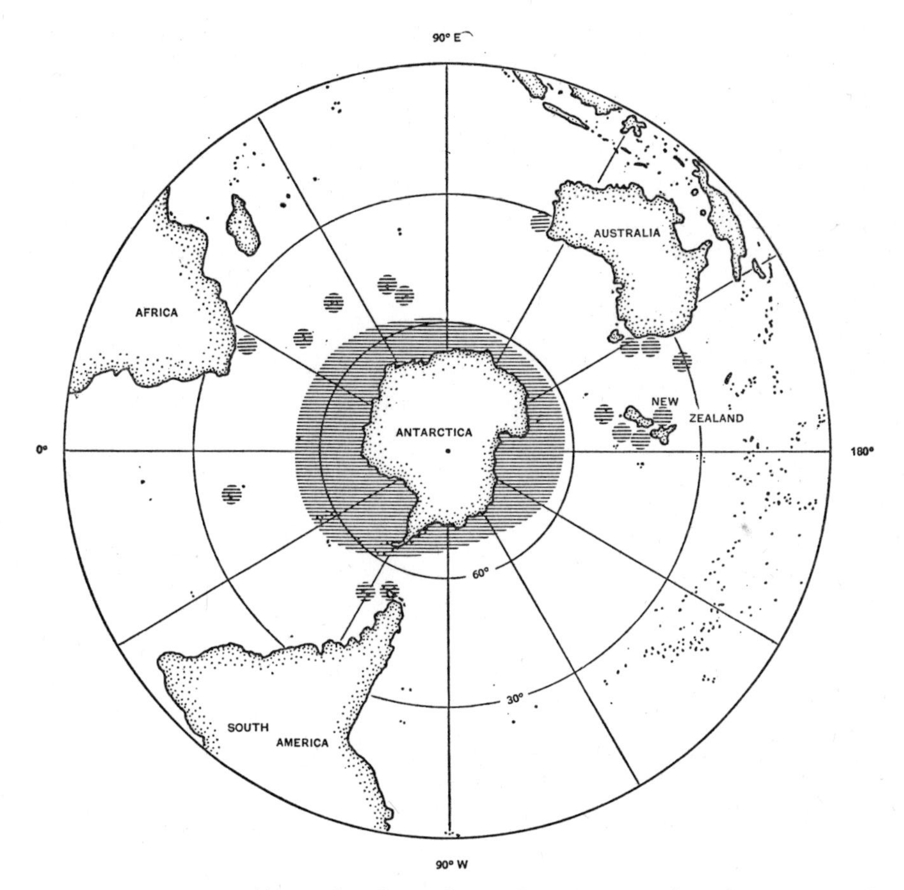

Fig. 11. Range of leopard seal. *Hydrurga leptonyx,* main breeding range and limital records.

South Georgia, year around in small numbers (Matthews, 1929, 1952); South Orkney Islands, "plentiful" (*Antarctic Pilot,* 1948, p. 131); Paulet Island, November to March (Andersson, 1908, p. 12); Deception Island (*Antarctic Pilot,* p. 165); Palmer Peninsula (= Graham Land), most numerous of the seals on the pack ice (Allen, 1905, p. 88); Falkland Islands, many records, mostly in spring and early summer (Hamilton, 1939*a*). *Ross Sea*: British Museum (Natural History) has specimens. *Victoria Land*: Most numerous of the seals on the pack ice (Allen, 1905, p. 88). *Campbell Island*: Reported here (Wilson, 1907, p. 26). *Macquarie Island*: Common from July to November (Ainsworth, 1934, p. 346; Gwynn, 1953*a*). *Adélie Coast*: Only 3 seen in 1949–51 (Sapin-Jaloustre, 1953, pp. 13–14). *Amsterdam Island*: A yearling seen in spring of 1955 (Paulian, 1957*b*, p. 211). *Heard Island*: Year around, winter concentration of about 900 in July–September (Gwynn, 1953*a*; Law and Burstall, 1953, p. 14; Brown, 1957, p. 4). Gwynn stated that the leopard seal is "primarily an animal of the outer fringes of the pack-ice . . . it is on the pack-ice that the young are born . . . it is possible that at Heard Island leopard seals are to be seen in greater numbers than anywhere else in the world." At Atlas Cove (53° 01′ S, 73° 22′ E) in late August 1951 he counted up to 85 seals in one group. Howard (1954) reported that 129 leopard seals had been hot-iron-branded through 1952. *Kerguelen Islands*: Breeding, seen from May to November, rarely as many as 3 together, taken once on Île Howe (Paulian, 1952, 1953, pp. 120–21; Angot, 1954, p. 11). *Crozet Islands*: "visits the islands" (*Antarctic Pilot*, p. 235). *Prince Edward Islands*: "the leopard seal probably occurs in limited numbers" (*Antarctic Pilot*, p. 230). *Bouvet Island*: The *Norvegica* collected specimens here in 1928–29 (Sivertsen, 1954, p. 54).

Limital records: *Tristan da Cunha*: One record (Barrow, 1910). However, "large spotted seals of fierce disposition are said to have been met with very occasionally at Cave Point and elsewhere" (Elliott, 1953, p. 42). *South America*: Cape Horn region (Wilson, 1907, p. 26; Cabrera and Yepes, 1940, p. 183). *New Zealand*: Port Nicholson, Wellington Harbour, Waikato River, and Wanganui River (Wilson, 1907, p. 26). Lord Howe Island (31° 31′ S), a specimen killed (Hamilton, 1939*a*, p. 242). *Australia*: Rare, New South Wales (Wilson, 1907, p. 26; Paulian, 1953, p. 118); Victoria (Brazenor, 1950, p. 72). "In 1921, after heavy gales lasting over a fornight, five strandings were recorded, and presumably there may have been many more on uninhabited stretches of the coast. The most northerly stranding . . . was at the Hawkesbury River near Sydney, and . . . a ten-foot male was captured in 1859 near the mouth of the Shoalhaven River, south of Sydney, with a full-grown platypus in its stomach" (Troughton, 1951, p. 259). Western Australia, first record, an adult which briefly visited 1 mile south of Gairdner River mouth on 12 October 1948 (Serventy, 1948). *South Africa*: An 8-foot female, alive, with two "kinds of Tern in the stomach," 40 miles north of East London (Roberts, 1951, p. 214).

Genus **LEPTONYCHOTES** Gill, 1872

Leptonychotes Gill, 1872, p. 70.

Leptonyx Gray, 1837, p. 582. Preoccupied in Aves.

TYPE. *Otaria weddelli* Lesson.

Leptonychotes weddelli (Lesson) 1826. (Weddell seal)

Otaria weddelli Lesson, 1826, p. 437 (spelled "*weddeli*" on p. 438). South Orkney Islands, Southern Ocean.

Leptonychotes weddelli, Allen, 1880, p. 467.

Ogmorhinus (Leptonychotes) weddelli, Trouessart, 1897, p. 381; 1904, p. 284.

TYPE. None. Species based on Robert Jameson's description (pp. 22–23) and James Weddell's poor drawing (between pp. 22 and 23) of the "sea leopard" (Weddell, 1825). The drawing is captioned "Sea Leopard of South Orkney's." Six seals were killed by the Weddell party on Saddle Island (60° 38′ S, 44° 50′ W), South Orkney Islands, on 15 January 1823. "The drawing of one deposited in the Edinburgh Museum is annexed." Barrett-Hamilton (1902, pp. 17–19) regarded the skull of this specimen as the type. Allen (1905, p. 91) pointed out, however, that Lesson's description of *Otaria weddelli* was based entirely upon Weddell's book.

Eleven years later Gray (1837, p. 582) gave the first intelligible description of the Weddell seal, "based upon two specimens (skins, with skull) [skulls "*a*" and "*b*" in Gray's catalog of 1862, p. 142] sent home by Captain Fitzroy, R.N., from the river Santa Cruz, in about latitude 50° South, on the east coast of Patagonia" (Barrett-Hamilton, 1902, p. 19).

RANGE (fig. 12). The Weddell seal is rather solitary, or semigregarious, in groups of up to 40 at pupping time (September–October); resident; feeds on small fishes and squids; is most closely attached of the Lobodontini to the fast bay-ice along shore; rarely wanders as far north as Uruguay (35° S). "The Weddell seal is circumpolar, and its normal habitat is the inshore waters of the Antarctic continent and adjacent islands. It spends much time in the water, but emerges at intervals to lie out on the beaches or on fast-ice. It is not a seal of the Antarctic pack-ice and is rarely found on isolated floes or far from land . . . It is the most southerly ranging mammal, apart from man himself, and is the seal most specialized for life in high latitudes, where so much of the year must be spent beneath the ice" (Bertram, 1940, pp. 5–6). Perkins (1945, pp. 278–79) has described the "nightmare of pressure ice" and fantastic formations in the Bay of Whales. Evidently the Weddell seals "remain below the ice, resting on interior ice shelves and probably making use of the anticlinal domes as places to breathe, for they can be heard calling beneath the ice all winter." He found three adult seals, branded by Lindsey 5–7 years earlier, "within a very short distance of the point where they were branded."

Principal records (east to west from the meridian of Greenwich): *Falkland Islands Dependencies*: South Shetland Islands (Bertram, 1940, p. 5). South Georgia, small permanent breeding colony (Matthews, 1929, 1952). South Orkney Islands, type locality, abundant, about 350 seen in 1933 (Bertram, p. 5; *Antarctic Pilot*, 1948, pp. 130–31). South Shetland Islands, a few breeding (Bertram, p. 5; *Antarctic Pilot*, p. 144). Deception Island, *Norvegica* specimens (Sivertsen, 1954, p. 55). Palmer Peninsula (= Graham Land), the site of Bertram's study. In all of the Depend-

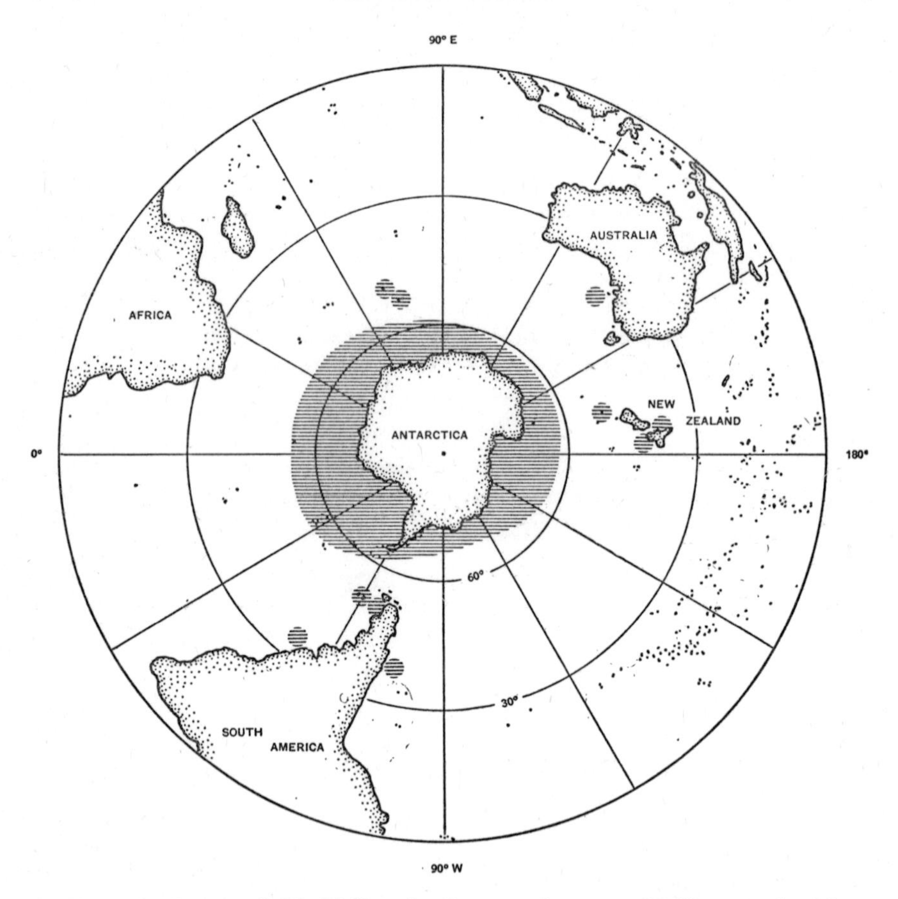

Fig. 12. Range of Weddell seal. *Leptonychotes weddelli,* main breeding range and limital records.

encies, Markowski (1952) listed 8 localities where cestodes were collected from Weddell seals; Laws (1953*a*) estimated a total of 800,000 Weddell seals. *Ross Sea and Victoria Land*: Breednig in large numbers, apparently the most common species of seal (Wilson, 1907, p. 69); seldom seen in the 4 months of darkness but certainly present beneath the ice; appearing in Bay of Whales about October (Lindsey, 1937). *Adélie Coast*: Hundreds in summer (Sapin-Jaloustre, 1952, p. 182). *Heard Island*: Rare, solitary; 1 pup seen in November 1950; "extreme northern limit of the distribution [53° S]" (Law and Burstall, 1953, p. 15). *Kerguelen Islands*: 49° N, reported by Wilson (1907). *Kemp Coast*: William Scoresby Bay; "numerous Weddell seals on 27 February 1936" (*Antarctic Pilot,* 1948, p. 301).

Limital records: *Antarctica*: Weddell seals probably reach 80° S at the end of the Bay of Whales (Lindsey, 1937). Wilson (1907, p. 20) reported occasional dead and sick animals more than 30 km. inland and at an elevation of 730 m. *Falkland Islands*: In 24 years Hamilton (1945) learned of only one record, at Cape Pembroke (51° 41′ S, 57° 42′ W), East Falkland Island, 28 October 1943. *Uruguay*: Isla de Lobos (35° S), perhaps the northernmost record (Vaz Ferreira, 1956*a*). *Patagonia*: Specimens from Río Santa Cruz (50° S) (Gray, 1837, p. 582); "common off the coast at Corriken Aike [= Departamento de Güer Aike]" (Hatcher, in Allen, 1905, p. 90). *Islas Juan Fernández*: Isla Más a Tierra (34° S); rare; records doubtful (Osgood, 1943, p. 98). *Chile*: Isla Mocha (38° S); seen every few years (Albert, in Allen, 1905, p. 90). *New Zealand*: Four occurrences, all in winter or early spring: Titahi Bay (18 km. north of Wellington), Wellington Harbour, Auckland Islands (Enderby Island), and Muriwai (32 km. west of Auckland) (Turbott, 1949). A specimen from Wanganui reported by Wilson (1907, pp. 12–13) was later shown to be a crabeater seal. *Australia*: A "straggler to South Australia" according to Iredale and Troughton (1934, p. 88); "a specimen was once taken along the coast of southern Australia" (Carter *et al.*, 1945, p. 105). *South Africa*: No records.

Subfamily CYSTOPHORINAE Gill, 1866

Cystophorinae Gill, 1866*a*, pp. 6, 9. (hooded seal and elephant seals)
Cystophorina Gray, 1837, p. 582. With the same meaning but different spelling.

REMARKS. Seals of the Cystophorinae are conspicuously marked by a secondary sex character: a large inflatible or erectile snout in the adult male, absent in the female and young male. Other features, such as reduction in number of lower incisors to one on each side of the jaw, suggest common ancestry of the two included genera. The population centers of the two are nearly poles apart at the present time. Numerous records show that individuals of both genera are prone to wander (*e.g.*, hooded seals in Florida and the Yukon; elephant seals in Alaska and the Island of St. Helena).

Genus **CYSTOPHORA** Nilsson, 1820

Cystophora Nilsson, 1820, p. 382.

TYPE. *Cystophora borealis* Nilsson (= *Phoca cristata* Erxleben).

REMARKS. Brass (1911, pp. 668, 671) refers to "*Cystophoca*" and "*Chrisophoca*"; quite certainly printer's errors, of which the book contains many.

Cystophora cristata (Erxleben) 1777. (hooded seal)
Phoca cristata Erxleben, 1777, p. 590. Southern Greenland and Newfoundland.

Cystophora borealis Nilsson, 1820, p. 383. Southern Greenland and Newfoundland.

Cystophora cristata, Nilsson, 1841, p. 326.

TYPE. None. Species based mainly on Pennant's (1771) "hooded seal" and Schreber's (1776) "Klappmüze." For history of names see Allen (1880, pp. 724–40). Locality, "in Groenlandia australiori et Newfoundland."

RANGE (fig. 13). The hooded seal is commonly associated with thick

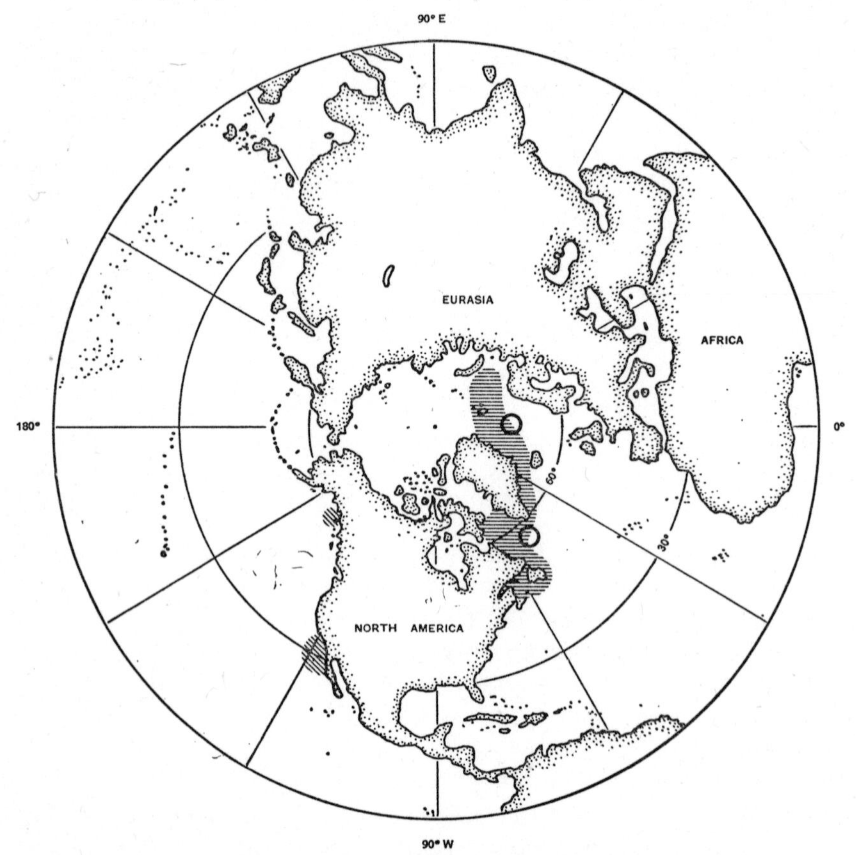

Fig. 13. Ranges of hooded seal and northern elephant seal. (≡) Hooded seal *Cystophora cristata.* (O) Breeding sites of the two main hooded seal stocks: Jan Mayen and Newfoundland. (\\\\\\) Northern elephant seal *Mirounga angustirostris,* main breeding range off California-Mexico and northernmost record, Alaska.

ice in deep waters of the Atlantic-Arctic; in small numbers except at breeding and molting times; migratory, nomadic. The reproductive unit is the family consisting of male, female, and single young per annum. (The closely related elephant seal, however, is strongly polygynous.) Lillie (1956, p. 62) stated that "as far as known the main concentrations of Hood seals are from June to September along the south eastern coastline of Greenland . . . A Harp seal migration starts from the Baffin Land area southwards about October and is joined by Hood seals from East Greenland as they travel down the Labrador coast toward the Grand Banks of Newfoundland. The Hoods tend to keep farther out to sea than the Harps, while waiting for the pack ice coming down behind them on the Greenland current. There are Hood concentrations also to the North of Yan Mayen but whether these also are migrants from the East Greenland area is not certain. Like the Harp seals, the Hoods in the Newfoundland sector haul out on the ice floes to give birth to the pups about the end of February."

Principal range (east to west; authorities same as for *Erignathus barbatus*, p. 109): Northern Siberia, mouth of Yenesei (about 80° E, accidental), Novaya Zemlya (rare), Barents Sea (a few, not every year), Bear Island, Svalbard (migrants on north coast in winter), Jan Mayen, Iceland (fairly common), Denmark Strait, southern Greenland, Baffin Island, Labrador, Newfoundland, Gulf of St. Lawrence, Hudson Strait (rare), Canadian Arctic (rare), near mouth of Mackenzie River (accidental at Tuktoyaktuk, 133° W, and Herschel Island, 138° W). "The Bladdernose does not occur in Hudson Bay and is not common on the north and east coast of Baffin Land" (Freuchen, 1935, p. 231).

Limital range (rare or accidental, north to south): From about 80° N in Siberian Arctic and 79° N in Canadian Arctic (Cape Sabine, Ellesmere Island, according to Anderson, 1947, p. 80), southward to Norway, British Isles, France (Gijzen, 1956, p. 22), and Portugal (Ellerman and Morrison-Scott, 1951, p. 333). Leon B. Strout photographed an adult female, and a pup in "blueback" stage, at North Harpswell, Maine, 25 March 1928. The pup was certainly the "newly born pup" listed by Norton (1930, p. 55) as a harbor seal. There are three records of hooded seals south of Cape Cod, the southernmost at Cape Canaveral, Florida (28° 27′ N; Miller, 1917; Goodwin, 1954). This last was an immature female.

Migration: The migration of the hooded seal is poorly understood. There seem to be two main breeding places: (1) The Jan Mayen site, or Norwegian Sea between Bear Island and Iceland, where hooded seals breed in spring and then disperse northward toward Svalbard in summer and fall. (2) The Newfoundland site south of Greenland. Here hooded seals breed in spring farther out to sea, and to the east of the harp seals. In summer and fall the hoods move to ice off Greenland. There seem to be complex local migrations for purposes of feeding and molting. There are, for example, concentrations of molting seals in Denmark Strait, between Iceland and Greenland, in May and June.

Isachsen (1933) reported an average annual take of 38,000 hooded seals off eastern Greenland in 1924–30. Backer (1948) saw about 5,000 at one time on the Jan Mayen site (73° N, 9° W).

Lillie (*op. cit.*) stated, ". . . it would seem optimistic enough to assume that 100,000 is the figure for Hood seals now left alive (exclusive of Russian territory) representing perhaps just a third of the original stocks in the combined areas of Greenland and Yan Mayen, and Newfoundland." But Sivertsen (*op. cit.*, p. 17) "estimated that the number . . . quoted by Dr. Lillie was much below reality." Zoologists of the North Atlantic region are in general agreement that, while the Jan Mayen population is not immediately threatened, the Newfoundland stock has been reduced to a few thousands and is facing complete extinction. Rasmussen (1957) has prepared an authoritative summary of current research on the harp and hooded seal populations of Denmark Strait and vicinity.

Genus **MIROUNGA** Gray, 1827

Mirounga Gray, 1827, p. 179. (elephant seals)

Macrorhinus É. Geoffroy Saint-Hilaire and F. Cuvier, 1826, p. 552. Preoccupied in Insecta.

Morunga Gray, 1843, pp. xxiii and 103. Spelling changed and name raised from subgeneric to generic level.

Macrorhinus, Trouessart, 1897, p. 377.

Mirounga, Allen, 1905, p. 94. Gray, 1827, credited as author; detailed history of name.

TYPE. *Phoca proboscidea* Péron (= *Phoca leonina* Linnaeus).

REMARKS. The elephant seals of the Northern Hemisphere (*M. angustirostris*) and those of the Southern (*M. leonina*) are regarded as barely distinct species. Before the advent of destructive commercial sealing in the nineteenth century the two groups were separated geographically by about 4,000 km., the distance between latitude 20° S on the Chilean coast (Bartholomew *et al.*, 1911) and 25° N on the Mexican coast. The species *M. angustirostris* was originally split from *M. leonina* upon comparison of the skull of a "probably full grown female northern elephant seal with the figure of a two thirds grown male skull of the southern form" (Laws, 1953*b*, p. 2). Through long experience with elephant seals, Laws was able to judge the age and sex (given as female by Gray, 1844) of the southern skull. Gill (1866*b*) gave few details of the differences supposed to exist, though he called attention to the "peculiarly narrowed and produced snout of the female [skull in *M. angustirostris*]."

Murphy (1914, p. 75) studied elephant seals on South Georgia where, he admitted, there were few really old males left after many years of exploitation. He concluded that "the snout of *Mirounga leonina* is entirely different from that of *M. angustirostris* . . . The whole nasal tube is narrower and shorter in the southern species, and is only slightly pendulous even in . . . the largest and oldest males. Nine out of ten of all those I saw at South Georgia had practically no 'trunks' at all. The face in profile reminded me of that of a rat . . ." (As a matter of fact, in the old bull northern elephant seal the snout is long and slender and hangs below the level of the chin.)

The northern and southern elephant seals seem to be about the same size. Thus the recorded maximum lengths are, for *males*, northern: 22 feet "from tip to tip" (presumably including hind flippers) (Scammon, in Allen, 1880, p. 743); *males*, southern: 21 feet 4 inches (Murphy, 1914, p. 74) and over 20 feet (Laws, 1953*b*, pp. 40, 52). For *females*, northern: 11 feet 5 inches, snout to tip of hind flippers (Rothschild, 1908, p. 394); *females*, southern: 11 feet 6 inches, snout to tail along curves of back (Laws, 1953*b*, p. 42).

Laws (1956*a*, p. 2) concluded that "in many respects the behaviour of *leonina* appears to be identical with that of *angustirostris* but the two species live under very different conditions . . . The southern species occasionally breeds on sea ice . . . and is able to keep open breathing holes in the ice for several weeks," whereas the northern species lives where midwinter seawater temperatures are above 13° C. (55° F.). Laws also observed (pp. 23–24) that the elephant seal on South Georgia is more excitable than is the northern form. On the Mexican seal, Bartholomew (1952) was actually able to lie prone! Bartholomew has pointed out (*in lit.*) that the northern elephant seal is distinctive not only in shape of proboscis but also in posture. "The northern species rarely, if ever, assumes the extreme U-shaped posture which is so characteristic . . . of the southern species" (see Laws, 1956*a*, p. 73 and pl. 4).

In summary, the northern and southern forms are widely separated in latitude and in climate, are different in profile of the snout and body, and are somewhat different in behavior. They are consequently regarded by most recent writers (including the present) as specifically distinct.

Mirounga leonina (Linnaeus) 1758. (southern elephant seal)

Phoca leonina Linnaeus, 1758, p. 37. Isla Más a Tierra, Islas Juan Fernández, Chile.

Phoca proboscidea Péron, 1816, vol. 2, p. 34, pl. 32. King Island, New South Wales, Australia. Type of *Mirounga* Gray.

Macrorhinus leoninus, Trouessart, 1897, p. 377; 1904, p. 282.

Mirounga leonina, Allen, 1905, p. 95.

Macrorhinus leoninus falclandicus Lydekker, 1909, p. 603, fig. 183. Falkland Islands.

Macrorhinus leoninus macquariensis Lydekker, 1909, p. 603, fig. 184. Macquarie Island and (?) Chatham Islands, New Zealand.

Macrorhinus leoninus crosetensis Lydekker, 1909, p. 606, fig. 185. Crozet Islands and (?) Kerguelen Islands and Heard Island.

TYPE. None. Flower (1884, p. 217) described, and Hamilton (1940) figured, fragments of a skull, Museum of the Royal College of Surgeons of England, Osteological Collection no. 3923; now held in British Museum (Natural History), no. 1946.8.9.1. These fragments were brought from Isla Más a Tierra by Lord Anson in 1744. There is no reason to believe, however, that Linnaeus saw this skull. His *Phoca leonina* was apparently based entirely on the description by Anson (1748, pp. 122–24) of the "Sea-Lyon," and a plate (between pp. 122 and 123) captioned "A Sea-Lion and Lioness" on Isla Más a Tierra, June 1741. Hamilton has proposed that the skull fragment be regarded as the type since it "is a contemporary of the type description, having indeed been collected when the type-material was examined." Linnaeus gave the range of the elephant seal as "Habitat ad polum Antarcticum."

RANGE (fig. 14). *General*: The southern elephant seal is circumpolar in subantarctic waters; casual from Saint Helena (16° S) to antarctic pack ice (78° S). It breeds from Gough Island and Argentina (about 42° S) to South Shetland Islands (62° S). It is thought to be migratory, moving in winter in various directions toward pelagic feeding grounds at the edge of the pack ice.

Entire range (east to west from the meridian of Greenwich): *Saint Helena*: 16° S, farthest north, no recent records, certainly rare (Fraser, 1935). *Gough Island*: 42° S, about 3,000 km. from the nearest continent, fewer than 300 elephant seals (Swales, in Holdgate *et al.*, 1956). *Tristan da Cunha*: Reported by Elliott (1953), Paulian (1953), and Laws (1956). Elliott stated that elephant seals formerly bred on Tristan Island and Inaccessible Island; young, supposedly from the Gough Island rookery, are seen frequently on Tristan beaches. *Falkland Islands Dependencies*: South Sandwich Islands: present and perhaps breeding (Matthews, 1952; Laws, 1956*a*). South Georgia: over 5,000 killed in some seasons (Matthews, 1952); not present the year around (Paulian, 1952); 259,000 exclusive of pups estimated in 1951 (Laws, 1953*a*). South Orkney Islands and South Shetland Islands: a breeding population of about 250 each and a summer visiting population, from South Georgia, of 10,000. The most southerly breeding grounds of *M. leonina* are Signy Island (60° 43′ S) in the South Orkneys and King George Island (62° S) in the South Shetlands (Laws, 1953*a*, 1956*a*). In the Falkland Islands and Falkland Islands Dependencies the total population has been estimated at 290,000 (Laws, 1953*a*). *Uruguay*: As far north as Isla de Lobos (35° S); reported in Río

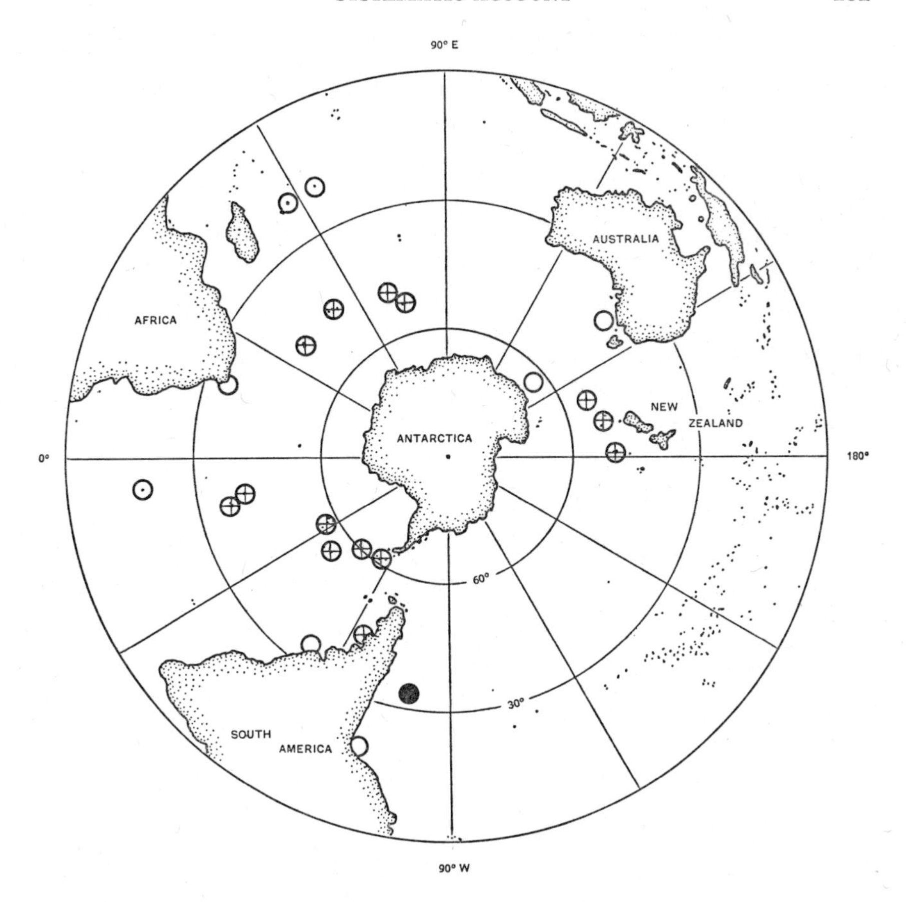

Fig. 14. Range of southern elephant seal, *Mirounga leonina*. (⊕) Breeding colonies. (○) Stray individuals or nonbreeding groups. (●) Type locality, Islas Juan Fernández, where seals were exterminated.

de la Plata up to 300 km. from its mouth (Vaz Ferreira, 1956*a* and *in lit.*). *Argentina*: Only one breeding colony ("apostadero") on Atlantic coast, at Punta Norte, Península Valdes, 42° 20′ S (González Ruiz, 1955, p. 123). Santiago Carrara (1952, appendix p. 4) estimated 115 elephant seals here. In 1954 he estimated 220. *Chile*: ". . . seems wholly extirpated" (Osgood, 1943). There seem to be no recent records from the type locality, Islas Juan Fernández. Bartholomew *et al.*, as late as 1911 (plate 4, map 5), showed "Phocidae" ranging up the west coast of Chile to about Iquique (20° S) and including Islas Juan Fernández but not the Isla San Ambrosio–Isla San Félix group. Heller (1904) did not mention elephant seals, past or present, on the Galapagos Islands. Cabrera and Yepes (1940, p. 186) stated that "as early as 1850 it was rare in Chile and Juan Fernández, where it has now disappeared." Murphy (1920, p. 94) stated that "an example has . . . recently been reported from Ilo [17° 40′ S] Peru." *Peter I Island, Bellingshausen Sea*: No suitable beaches, hence one of the few antarctic islands without elephant seals (*Antarctic Pilot*, 1948, p. 71). *New Zealand*: Antipodes Islands, Campbell Island, Auckland Islands, Macquarie Islands: reported by Doutch (1952, p. 226), Matthews (1952), and Laws (1956*a*). Mielche (1953) visited Campbell Island in early January 1951 "just after the breeding season" and collected six specimens. Sorensen (1950) counted 417 seals here in 1947; he thought the Campbell seals were an overflow from Macquarie. Doutch counted 70,000 on Macquarie Islands in December 1950. "On a recent visit to the Antipodes Islands, several small harems were seen" (Turbott, 1952, pp. 199–200). *Australia*: Formerly resident in Bass Strait, now sporadic (Iredale and Troughton, 1934, p. 88; Laws, 1956*a*). *Adélie Coast*: A young male killed on 21 January 1951; no others seen (Sapin-Jaloustre, 1953). This is on or near the Antarctic Circle. *Amsterdam Island and St. Paul Island*: Reported by Laws (1956*b*); said to be rare on Amsterdam (Paulian, 1953). *Heard Island*: About 40,000 in 1951; height of breeding season late October (Law and Burstall, 1953). Over 2,000 have been hot-iron-branded in recent years (Howard, 1954). There seem to be no records from McDonald Island but presumably elephant seals haul out here. *Kerguelen Islands* (49° S): Present throughout the year; height of pupping season first of October; height of mating season first of November (Paulian, 1952; Angot, 1954). Later, Paulian (1957*b*, p. 220) stated that the Kerguelen population is clearly less than 100,000. *Mauritius and Rodrigues Island*: Several records of young animals as far north as 19° 40′ S (Vinson, 1956). *Crozet Islands*: Reported by Matthews (1952) and Laws (1956*a*). *Prince Edward Islands,* including Marion Island: 1,500 killed in 1933 (Rand, 1956*a*). *South Africa*: Cow with newborn pup on Cape Agulhas, 35° S, about 2,000 km. from the nearest rookeries on Prince Edward Islands (Kettlewell and Rand, 1955). A large male elephant seal came ashore at Cape Town during a storm and remained on the rocks for two days (Anonymous, 1949). *Bouvet Island*: "No mammals were observed by the 'Valdivia' or the 'Norvegica,' but seals are reported to have been taken on the island and probably still frequent it in small numbers. They are most likely to be elephant, and possibly leopard seals" (*Antarctic Pilot*, 1948, pp. 229–30). However, in the same book (p. 71) Roberts stated that Bouvet Island has no beaches suitable for elephant seals.

REMARKS. The three subspecies of *M. leonina* proposed by Lydekker may or may not be valid. It is known that elephant seals continue to grow and change in size and shape for more than 20 years. They also wander widely and may (?) intermingle. Willett (1943) gave a record of a nomad

3,000 km. from home, and Kettlewell and Rand (1955) a record of another at least 2,000 km. from home.

Mirounga angustirostris (Gill) 1866. (northern elephant seal)

Macrorhinus angustirostris Gill, 1866*a,* p. 13 (separates issued 7 April); 1866*b,* p. 13 (April). Dates according to Poole and Schantz (1942, p. 234). Bahía Tórtola, Baja California, Mexico.

Macrorhinus angustirostris, Trouessart, 1897, p. 377; 1904, p. 282.

Mirounga angustirostris, Elliot, 1904, p. 545.

Mirounga leonina leonina, Rothschild, 1910, p. 446. The Mexican form regarded as the typical one, representatives of which were wintering on Isla Más a Tierra when discovered by Lord Anson in 1741. There is, however, no evidence whatsoever that elephant seals migrate through equatorial waters.

TYPE. *Lectotype*: A female skull "twelve and a half inches in length" regarded by Laws (1953*b,* p. 2) as probably full-grown, U.S. National Museum no. 4704, collected in Bahía Tórtola (= Bahía San Bartolomé) 27° 39′ N, 114° 51′ W, Baja California, Mexico, in 1857, by W. O. Ayres. "Type not designated by original describer. This is the only specimen in the Museum answering to the locality and dimensions published by T. N. Gill. It was figured by J. A. Allen [1880, figs. 57–60], who designated it as the type (= lectotype) on p. 748, footnote to table" (Poole and Schantz, 1942, p. 234).

RANGE (fig. 13). The northern elephant seal is gregarious and resident (?), though individuals may wander far. It probably feeds in deep water. One taken 40 miles from shore had eaten a number of small sharks, squids, and rays. The fact that well-marked ridges appear on the roots of the teeth of an adult female from Bahía San Cristobal (U.S. National Museum no. 21889) indicates seasonal interruption in the feeding schedule. This in turn suggests a long fast, but whether migration away from land is involved cannot be said with assurance.

The elephant seal formerly ranged from Point Reyes, California (38° N), to Cabo San Lázaro (near Bahía Magdalena), Baja California (24° 48′ N), a distance of about 1,700 km. As a result of destructive commercial sealing it was nearly exterminated—perhaps reduced to a few score individuals—by 1892. Today there are 8,000 to 10,000 elephant seals distributed as follows: *California*: San Miguel Island (34° 04′ N, occurring regularly); San Nicolas Island (33° 15′ N, several counts of over 100 in spring, breeding?). *Baja California, Mexico*: Islas los Coronados (breeding on the south island, at least); Isla San Jerónimo (occurring regularly); Isla de Guadalupe (250 km. off the mainland, several breeding colonies, 4,548 elephant seals estimated in January–February 1950); Islas San Benito (28° 18′ N, at least 908 seen in January–February 1950 on the east, middle, and west islands, breeding in spring). The foregoing data are largely from Bartholomew (1952, p. 371; 1955, p. 6), Bartholomew and Hubbs (1952), and Berdegué (1956). The present writer visited Isla de Guadalupe in June 1955.

North of the breeding range individual elephant seals are seen fairly regularly by offshore fishermen as far north as Vancouver Island, Canada. At least two specimens have been taken in British Columbia waters (Cowan and Guiguet, 1956, p. 354). A far northern record was set by a subadult male that stranded, dead but fresh, at Kasaan Bay (55° 30′ N, 132° 25′ W), Prince of Wales Island, Alaska, at least 3,000 km. north of the breeding ground (Willett, 1943).

6

SYNOPTIC KEY TO THE GENERA

As already stated, two main groups of pinnipeds are recognizable: a walking group with swimming mechanism centered near the fore body and a wriggling group with swimming mechanism centered near the hind body. Associated with these differences in locomotor performance are many structural features of diagnostic value at the highest (superfamily) level. Other features, not associated with locomotion, are shared by members of both superfamilies and are thus of diagnostic value only at familial or generic levels. Examples of the latter are: presence of external ears, presence of scrotum, presence of sagittal crest, number of mammary teats, and number and arrangement of teeth.

In connection with teeth it should be cautioned that dental formulae are not always diagnostic. The number of teeth in pinnipeds is perhaps more inconstant than in land carnivores. For example, Allen (1880, p. 731) figured an extra pair of lower incisors in a hooded seal and "Dr. Nehring found in several examples of *Halichoerus grypus* the normal five back teeth increased to six" (Beddard, 1902, pp. 447–48). Walrus skulls with three upper canines (tusks) are fairly common. Unless otherwise noted below the characters given in the key are those of full-grown individuals, though many characters of course are present from birth. Certain anatomical terms are explained (in parentheses) for greater clarity. In the absence of an atlas of pinniped anatomy the terminology of M. E. Miller's *Guide to the Dissection of the Dog* (1952) has been found useful. Terminology of measurements: *body length,* on unskinned body stretched to its fullest, a straight line from tip of nose to tip of tail. *Body weight,* of entire animal. *Condylobasal length* (CBL), from transverse line touching most posterior points on occipital condyles, to transverse line touching most anterior points on premaxillary (= intermaxillary) bones (see fig. 15). *Basilar length of Hensel* (BLH), from most posterior point at middle of inferior lip of foramen magnum, to transverse line touching posterior margins of alveoli of median incisor teeth. *Mastoid width* (MW), greatest width between mastoid processes, outside of one to outside of the other. *Zygomatic width* (ZW), greatest width, at right angles to axis of skull, between zygomatic arches, outside of one to outside of the other.

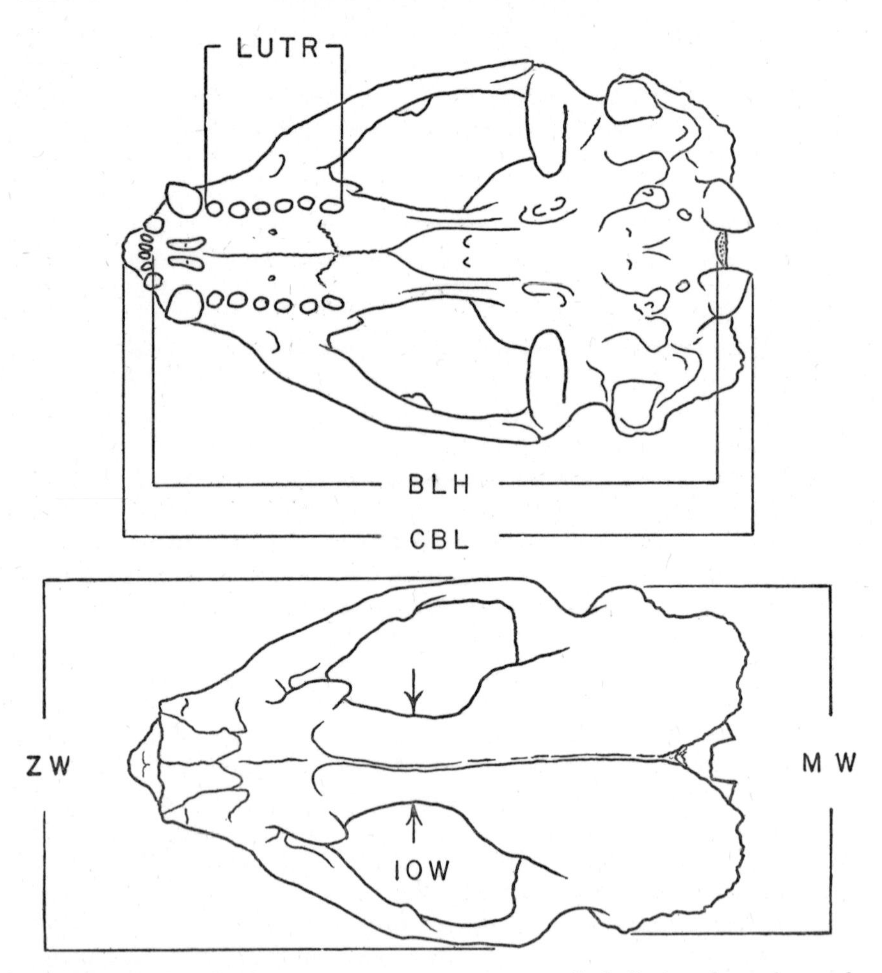

Fig. 15. Standard measurements on a pinniped skull (northern fur seal, adult male): BLH = basilar length of Hensel, CBL = condylobasal length, IOW = interorbital width, LUTR = length of upper tooth row, MW = mastoid width, ZW = zygomatic width.

Interorbital width (IOW), least distance across top of skull above and behind orbits. (This equals "interorbital constriction" of Hall, 1946, p. 679; not "interorbital constriction" of Sivertsen, 1954, fig. 5. It would be represented on Sivertsen's figure by a dotted line midway between "7" and "8".) *Length of upper tooth row* (LUTR), on axis of skull, from transverse line touching posterior margins of last pair of molars to transverse line touching anterior margins of first pair of incisors.

As an expression of the full-grown size attained by males and females in each genus, certain data on body length, body weight, CBL, and MW have been selected from published records and the writer's notes. The "size" concept is composite. For example, the values for body length and for body weight are not necessarily those of one individual; each is the largest value of its kind known to the writer. Occasionally an estimate (preceded by *circa*) is given where no actual value is known. Future studies will quite certainly show that some of the values given here are low.

Key

1 Locomotion on land accomplished by walking or running movements; hind limbs can be turned forward to support part of weight of body. Locomotion in water (swimming and diving) accomplished mainly by use of fore limbs; area of fore flippers great. Ankle joint resembles that of land carnivores; distinct groove on astragalus to fit end of tibia (astragalus is principal ankle bone, on same side of flipper as first digit); astragalus without calcanear process; femur with small trochanter minor (process on proximal end of thigh bone). Fore flippers long and oarlike; more than one-quarter length of body (otariids) or about one-quarter (walrus); flipper thick and cartilaginous, distinctly thicker at forward (leading) edge; digits stiff, with softer cartilaginous tips extending far beyond claws. Three distinct claws always present on middle digits of hind flipper (those on marginal digits always rudimentary); cartilaginous extensions well beyond all claws; first digit (big toe) of hind flipper nearly equal to, or shorter than, middle three digits. Scapula (shoulder blade) roughly triangular (otariids) or squarish (walrus), width considerably more than one-half the length. Vertebral spines of anterior thorax moderately developed (limiting upward flexibility of neck). Tympanic bone relatively small, flattish, thin-walled; mastoid process often prominent, especially in males; alisphenoid canal present (internal maxillary artery is enclosed in a bony ridge which passes along the outer side of the base of the pterygoid process); pterygoid (Vidian) canal (through which the superficial petrosal nerve leaves the cranial cavity) present. Nasal bones touch, but do not form a distinct wedge between, frontal bones (impossible to see in adult walrus); external nares more anterolateral. Postcanines of most species with one prominent cusp, some with 1 or 2 minor cusps. Pelage colors rather uniform, never spotted or sharply banded; distinct underfur sometimes present in adult; newborn pelage hairy, never woolly; all surfaces of all flippers sparsely haired or naked; mystacial vibrissae smooth; superciliary vibrissae few and small. Mammae with 4 teats. All species polygnous; males distinctly larger than females. Breeding habitat exclusively marine............Otarioidea (Otariidae and Odobenidae), 2

1′ Locomotion on land accomplished by wriggling movements; hind limbs cannot be turned forward. Locomotion in water mainly by use of hind limbs; area of fore flippers moderate. Ankle joint specialized, ball-and-socket-like; astragalus smooth, with long calcanear process; femur without trochanter minor. Fore flipper shorter, much less than one-quarter length of body; flipper flexible and of nearly uniform thickness; tips of digits extend little if any beyond claws; claws distinctly functional, often large. Five claws on hind flipper, in some species rudimentary. Scapula roughly falcate (sickle shaped), width about one-half the length. Vertebral spines of anterior thorax less well developed, allowing greater upward flexibility. Tympanic bones large, inflated, thick-walled; mastoid processes swollen but not prominent; alisphenoid canal absent; pterygoid canal absent. Nasal bones form a wedge between frontals; external nares more dorsal. Postcanines of most species with 3 or more distinct cusps. Pelage often spotted, occasionally sharply banded; distinct underfur never present in adult; newborn pelage woolly in most species, hairy in a few; all surfaces of all flippers distinctly haired; mystacial vibrissae often beaded; superciliary vibrissae well developed. Mammae with 2 or 4 functional teats. Most species monogamous, with male equal to, slightly smaller, or slightly larger than female. Breeding habitat often estuarine (river mouth); some species permanently in lakes.....Phocoidea (Phocidae), 8

2(1) Body form slender and elongated. Articular facet on astragalus rounded, less distinct. Pes and manus relatively larger, upper arm shorter (emphasizing ability to swim). Scapula less like that of land carnivores, low and broad, triangular. Fifteen thoracic vertebrae and pairs of ribs. All surfaces of all flippers naked, smooth, and leathery from tip to base. Fore flipper with 5 claw rudiments in pits. External ear pinna with cartilage, pointed, conical, up to 50 mm. (nearly 2 in.) in length from notch. Small but distinct tail projecting beyond outline of body. Incisors of deciduous (fetal) dentition, on each side: $\frac{1\text{-}2\text{-}3}{0\text{-}2\text{-}3}$. Adult dentition generalized; back teeth may atrophy, though leaving at least 5 postcanines on each side. Upper canine normal, ceases to grow when body is full-grown; apical foramen closes in old age (about 15 years); 1st and 2d upper incisors with transverse groove, 3d caniniform; postcanines simple, haplodont (though with rudimentary accessory cusps), never more than 1 lower but as many as 7 upper; the 5th upper postcanine usually, but not always, double-rooted. Surface of mastoid process not continuous with auditory bulla. Halves of lower jaw not solidly fused; more than one important mental foramen. Supraorbital process well developed; sagittal crest present, large in old males. Eye large; orbital opening of skull up to 76 mm. (3 in.) in diameter (*Eumetopias*). Anteromedial (inner) wall of orbit defective, perforate. Mystacial vibrissae less than 100 in number, up to 46 cm. (18 in.) in length; flexible, smooth. Tip of tongue notched. Testes scrotal. Never breeding in arctic waters. Young do not normally swim for at least a fortnight............................Otariidae, 3

2′ (See plates 14 and 15.) Body form thick and swollen. Articular facet on astragalus cylindroid, more distinct. Pes and manus relatively smaller, upper arm longer (emphasizing ability to support body weight on land). Scapula more like that of land carnivores, high and narrow; rectangular. Normally 14 thoracic vertebrae and 6 lumbar (but Turner, 1912, p. 164, described a skeleton with a 15th pair of very small ribs). Upper surface of all flippers sparsely haired (at least in young); lower surfaces naked. Fore flipper with 5 small but distinct claws. External ear pinna without cartilage, merely a low wrinkle of skin. No free tail. Incisors of deciduous (fetal) dentition, on each side: $\frac{1\text{-}2\text{-}3}{1\text{-}2\text{-}3}$. Adult dentition highly specialized for crushing molluscs; most of front and back teeth atrophy, leaving at least 3 cheek teeth on each side. All adult teeth except the canines become rather uniformly bluntly conical or flat; upper canine in both sexes continues to grow throughout life into an enormous tusk up to 100 cm. (39.5 in.) total length, or 79 cm. (31 in.) exposed portion, in male; 61 cm. (24 in.) total length in female. Apical foramen of canine remains widely open (as in some phocids). First and 2d upper incisors (present in young only) ungrooved. All postcanines single-rooted. Surface of mastoid process continuous with auditory bulla; muscles inserting on process cause it to grow into an enormous projection which expands beneath the outer meatus. (The mastoid process in the walrus was termed by Bobrinskoi the "mammiform" process.) Halves of lower jaw solidly fused in adult; single conspicuous mental foramen. Supraorbital process absent; sagittal crest absent. Eye smaller; orbital opening of skull up to 51 mm. (2 in.) in diameter. Anteromedial (inner) wall of orbit entire, not perforate. Mystacial vibrissae up to 400 in number, up to 30 cm. (11.8 in.) in length; stiff and thick as a crowquill; forming a conspicuous bouquet on each side of snout; smooth. Tip of tongue rounded. Testes internal, not scrotal. Breeding only in arctic waters. Young swim freely at birth.

Large males, length 366 cm. (144 in.); weight 1,268 kg. (2,795 lbs.); CBL 412 mm.; MW 342 mm. Large females, length 299 cm. (118 in.); weight 850 kg. (1,874 lbs.); CBL 339 mm.; MW 253 mm. (Nikulin, 1940, gave 392 cm. for males, 342 cm. for females; though "maximum length" may have included hind flippers.).......Odobenidae (*Odobenus*), p. 84

3(2) Pelage with one layer of coarse hair (though sparse, fine underhairs are present); newborn young brown to dark brown. First digit of fore flipper longer than 2d; marginal digits of hind flippers distinctly longer than middle 3. Snout more blunt. Maximum weight about 1,016 kg. (2,240 lbs.) in male. Occasionally ascending rivers (South American sea lion 300 km. up Río de la Plata; Steller sea lion 150 km. up Columbia River).

Note. Sivertsen's tables of measurements and photographs (1954, pp. 22–27, pls. 1–6) provide valuable aid to identification of skulls of adult male otariids. The genus *Arctophoca,* regarded as valid by him, is treated under *Arctocephalus* in the present review. King (1954, pp. 334–35) has prepared a key to the skulls (both sexes, mainly adult characters) of Pacific American otariids.................................Otariinae, 4

3′ Pelage with two distinct layers: dense, velvety underfur and coarse overhair; newborn young glossy black. First digit of fore flipper shorter than 2d; digits of hind flippers approximately equal in length. Snout more pointed. Maximum weight about 295 kg. (651 lbs.) in male. Rarely entering fresh water................................Arctocephalinae, 7

4(3) (See plate 1.) *Male.* Palate very long, extending nearly to pterygoid processes (hamuli); distance from palatal notch to incisors greater than 55 percent CBL; hind margin truncate, nearly a straight line transverse to long axis of skull; lateral margins raised, especially posteriorly, making the surface deeply concave, depth of concavity to 40 mm.; snout broad and upturned; width of skull immediately behind alveoli of canines usually more than 25 percent CBL; pterygoid processes long and able to support part of weight of skull resting on table. Temporal process present (a conspicuous process on the anterior bulge of the cranium at the level of the hamuli, almost as large in the old male as the supraorbital process; also present, though smaller, in adult female; apparently does not occur in other pinnipeds).

Large males, length 246.4 cm. (97 in.); weight *c.* 522 kg. (1,150 lbs.); CBL 393 mm.; MW 240 mm. Large females, length 195.6 cm. (77 in.); weight?; CBL 277 mm., MW *c.* 130 mm....................*Otaria,* p. 53

4′ *Male.* Palate shorter, distance from palatal notch to incisors less than 55 percent CBL; hind margin notched or deeply concave (except in *Eumetopias,* category 5); surface flat or slightly concave, depth of concavity never more than 10 mm.; snout narrower, width of skull immediately behind alveoli of canines less than 29 percent CBL; pterygoid processes shorter, not touching table. Temporal process absent. Old male with distinct whitish or yellowish cap..............................5

5(4) (See plate 2.) Conspicuous gap between 4th and 5th postcanines in both sexes, in male 5 to 8 percent CBL, or about the width of two teeth. Supraorbital process quadrate (squarish). Rarely more than 5 postcanines in each upper jaw.

Large males, length 315 cm. (124 in.); weight 1,016 kg. (2,240 lbs.); CBL 398 mm.; MW 232 mm. Large females, length 231 cm. (91 in.); weight 274 kg. (605 lbs.); CBL 327 mm.; MW *c.* 158 mm. (Rass *et al.* 1955, p. 99, gave slightly greater values: males 353 cm. and 1,120 kg.; females 270 cm. and 350 kg.) . . . . *Eumetopias,* p. 56

5′ Gap between 4th and 5th postcanines smaller or absent, in males 0 to 4 percent CBL. Supraorbital process triangular and pointed backward. Most, but not all, species with 6 postcanines in each upper jaw. Size smaller 6

6(5) (See plate 3.) *Both sexes.* Sagittal crest rising steeply from the frontal, its summit anterior to level of glenoid fossae, sloping evenly down to much lower occipital crest. Sagittal crest high, up to 38 mm. or 12 percent CBL. Voice a sharp, barking or honking "aarnh! aarnh! aarnh!" as against the more sustained roaring call of other otariids. *Male* (as further distinct from all other sea lions). Snout rather small, distance from palatal notch to incisors less than 45 percent CBL; width of skull at canines less than 23 percent CBL; width of anterior narial opening of skull less than 11 percent CBL. *Male* (as further distinct from *Neophoca,* category 6′). Anterior region of palate flat or only slightly concave; depth of concavity up to 9 mm. at level of canines. *Male* (as further distinct from *Arctocephalus,* category 7). Specimens from California more often with 5 than 6 upper postcanines on each side.

Large males, length *c.* 236 cm. (93 in.); weight *c.* 281 kg. (620 lbs.); CBL 330 mm.; MW *c.* 175 mm. Large females, length *c.* 183 cm. (72 in.); weight *c.* 91 kg. (200 lbs.); CBL 251 mm.; MW *c.* 110 mm. (Partly from Ray Gilmore, *in lit.*) . . . . *Zalophus,* p. 59

6′ (See plate 4.) *Both sexes.* Summit of sagittal crest posterior to level of glenoid fossae; height of crest up to 30 mm. or 10 percent CBL. *Male* (as further distinct from *Zalophus*). Snout generally larger, distance from palatal notch to incisors greater than 45 percent CBL, width of skull at canines greater than 23 percent CBL, width of anterior narial opening of skull greater than 11 percent CBL. Anterior region of palate deeply concave; depth of concavity at level of canines up to 18 mm.

Large males, length 236 mm. (93 in.); weight *c.* 408 kg. (900 lbs.); CBL 346 mm.; MW 181 mm. Large females, length ?; weight *c.* 227 kg. (500 lbs.); CBL 293 mm.; MW 126 mm. A poorly known genus. . . . . *Neophoca,* p. 64

7(3) (See plate 5.) Interorbital region short, usually less than 20 percent CBL (*cf.* Sivertsen, 1954, pls. 6–7); tympanic bulla convex. *Male.* Top of snout descends gradually, a line from nasals to gnathion makes an angle of less than 60° with long axis of skull; anterior narial opening of skull is less than 14 percent CBL. Snout long, palatal notch to incisors greater than 37 percent CBL; gnathion to posterior palatal end of maxilla greater than 44 percent CBL; sides of snout clearly depressed at level of 3d post-canine. *Male* (as further distinct from *Zalophus*). Skull more compact, 40 to 70 percent heavier (Sivertsen, 1954, p. 33); tends strongly to have 6 rather than 5 postcanines. *Both sexes* (*Arctocephalus philippii* only, as further distinct from its geographic neighbor *Callorhinus*). Nasals long and slender, combined width at anterior ends about 40 to 50 percent of their length. Sivertsen (1954, p. 49) stated that in most of the species of *Arctocephalus* the upper and lower postcanines have a marked cingulum and one or two more or less prominent additional cusps, except in *A. gazella,* which never has secondary cusps.

Large males, length 257 cm. (101 in.); weight 295 kg. (651 lbs.); CBL 286 mm.; MW 176 mm. Large females, length 179 cm. (70 in.); weight 122 kg. (268 lbs.); CBL 239 mm.; MW 110 mm. . . . . . . *Arctocephalus,* p. 67

7′ (See plates 6–13.) Interorbital region longer, usually greater than 20 percent CBL (*cf.* Sivertsen, 1954, pl. 5); tympanic bulla concave. *Male.* Top of snout projects forward, descending more abruptly, a line from nasals to gnathion making an angle of greater than 60° with long axis of skull; anterior narial opening of skull greater than 13 percent CBL; snout shorter, palatal notch to incisors less than 41 percent CBL; gnathion to posterior end of maxilla less than 45 percent CBL; sides of snout faintly depressed at level of 3d postcanines. *Both sexes* (as further distinct from geographic neighbor *Arctocephalus philippii* but not from *A. australis*). Nasals shorter and wider, combined width at anterior ends about 80 to 90 percent of their length.

Large males, length 214 cm. (84 in.); weight 278 kg. (613 lbs.); CBL 262 mm.; MW 147 mm. Large females, length 143 cm. (56 in.); weight 63 kg. (138 lbs.); CBL 199 mm.; MW 102 mm. . . . . *Callorhinus,* p. 82

8(1) Dentition more primitive; 3 upper incisors on each side; a 6th upper postcanine occasionally present (frequently in *Halichoerus*). Hind digits (toes) nearly equal in length. Anteromedial (inner) wall of orbit entire. Premaxillary (intermaxillary) bones prolonged upward, meeting the nasals. Representatives of all genera inhabiting arctic, subarctic, or temperate regions of the Northern Hemisphere (*Phoca* as far south as 28° 12′ N) . . . . *Phocinae,* 9

8′ Dentition less primitive; 2 upper incisors on each side (1st upper lacking); 6th upper postcanine not normally present. First and 5th hind digits (toes) clearly longer than middle 3. Anteromedial (inner) wall of orbit defective (or partly so in Cystophorinae). Premaxillary (intermaxillary) bones not reaching the nasals (except in *Leptonychotes*). Representatives of all but one genus (*Cystophora*) inhabit subtropical waters or Southern Hemisphere. 14

9(8) (See plate 23.) Mystacial vibrissae smooth, thick, straight, conspicuously bushy; their development related to bottom-feeding habit. Third digit of fore flipper longest. Jugal (malar) bone short and wide; width more than one-third its greatest diagonal length. No sagittal crest. Mammary teats 4. Embryonal (newborn) pelage dark grayish or brownish. Teeth of adult loosely rooted; often worn down to surface of jaw or lost entirely. Spaces between postcanines almost tooth-wide.

Large males, length 285 cm. (112 in.); weight 397 kg. (875 lbs.); CBL 230 mm.; MW 144 mm. Large females, length 260 cm. (102 in.); weight ?; CBL 220 mm.; MW 131 mm. (Weight of male is from Soper, 1944, p. 240; eviscerated animal [= 744 lbs.] plus 15 percent.) ....Erignathini (*Erignathus*), p. 109

9′ Mystacial vibrissae beaded, slender, curled, less bushy. Third digit of fore flipper shorter than 1st and 2d. Jugal (malar) bone longer and narrower, width less than one-third its greatest diagonal length. Sagittal crest present in some species. Mammary teats 2. Embryonal pelage (persistent at birth except in *Phoca*) white. Teeth of adult firmly rooted and functional. Spaces between postcanines less than tooth-wide, or absent. ....*Phocini,* 10

10(9) (See plate 22.) Snout long; distance between tip of nose and eye almost twice that between eye and ear; profile of forehead and snout straight or convex. Postcanines large and strong, some as wide (transverse to axis of jaw) as long (in axis of jaw); each usually with single conical cusp; secondary cusps, when present, on hind postcanines only and insignificant; 1st upper postcanine in old animals pushed medially out of line; 5th upper postcanine separated from 4th; 6 upper postcanines occasionally present. Fore limbs flexible, often used for crawling or climbing; fore claws long, slender, curved. Sagittal crest well developed on skull of old male. Width of anterior narial openings of skull more than 30 percent of MW. Crown pelage usually lighter than back; neck of adult male (as distinct from *Phoca*) obese, with 3 or 4 conspicuous wrinkles. Closed nostril slits separated by an inch or more, seen from the front as a \ / figure (as distinct from *Phoca,* in which the slits almost meet in a broad V). (Wynne-Edwards, 1954.) Polygynous; male distinctly larger than female. Breeding in fall or winter, mid-August to March.

Large males, length *c.* 300 cm. (118 in.); weight *c.* 290 kg. (640 lbs.); CBL 327 mm.; MW 158 mm. Large females, length *c.* 230 cm. (91 in.); weight *c.* 249 kg. (550 lbs.); CBL 262 mm.; MW 139 mm. . . . . *Halichoerus,* p. 106

10′ Snout short; distance between tip of nose and eye much less than twice that between eye and ear; profile of forehead and snout concave. Postcanines smaller, thinner, each with 2 or more cusps. Fore limb not especially modified for crawling. Sagittal crest moderately developed (or absent in some species). Width of anterior narial openings of skull less than 30 percent of MW. Crown pelage not lighter than back. Monogamous (promiscuous); sexes nearly equal in size or male slightly larger. Breeding in spring or summer 11

11(10) Posterior margin of palate distinctly notched or incised (Doutt, 1942, pl. 9). Posterior free margin of mesethmoid (bony nasal septum) falls far short of posterior edge of palate. Adult pelage usually spotted (rarely in *Pusa sibirica*); male and female adult pelage alike. Commonly breeding in harbors and fjords. 12

11′ Posterior margin of palate not distinctly notched or incised, nearly a straight line. Posterior free margin of mesethmoid reaches, or nearly reaches, posterior edge of palate. Adult pelage unspotted (except in certain females which have just reached maturity); marked with large bands of dark. Male and female pelage unlike. Commonly breeding on open sea ice. 13

12(11) (See plates 16–19.) Dentition heavier, adapted to diet of fish and shellfish. Maximum CBL more than 187 mm.; length (along axis of tooth row) of 2d upper postcanine usually 6.8 mm. or more; mandibular teeth often crowded out of line and overlapping; 1st lower postcanines usually with 4 cusps (the 2d from anterior end largest); inner side of mandible between middle postcanines convex; posterior palatine foramina usually enter palate anterior to maxillo-palatine suture. IOW greater than 7 mm. Claws more semicircular in cross-section, without distinct dorsal ridge or annuli (growth layers). Skull of old animals of both sexes with low sagittal crest; face more doglike, snout blunter. Old male without disagreeable odor. Pelage on back smooth to the touch, hairs finer, tips recurved. Spots small, often in clusters, not especially ring-shaped. (Bailey and Hendee, 1926, pl. 3, and Kenyon and Scheffer, 1955, p. 22, 25, illustrated the distinction between pelage spots of *Phoca* and *Pusa.*) Young born on land in early summer; grayish-white embryonal pelage normally shed before birth (except, apparently, in certain far northern strains).

Large males, length 173 cm. (68 in.); weight 116 kg. (256 lbs.); CBL 237 mm.; MW 138 mm. Large females, length 154 cm. (61 in.); weight 110 kg. (243 lbs.); CBL 222 mm.; MW 124 mm.............. *Phoca,* p. 88

12′ (See plate 20, lower.) Dentition weaker, adapted to a diet which may for long periods contain only macroplankton (Dunbar, 1949). Maximum CBL 187 mm. or less; length of 2d upper postcanine usually less than 6.8 mm.; mandibular teeth always aligned with jaw, never crowded; postcanines usually with 3 cusps (the center one longest); inner side of mandible between middle postcanines concave; posterior palatine foramina usually enter palate posterior to maxillo-palatine suture. IOW less than 7 mm. Claws more triangular in cross-section, with distinct dorsal ridge and annuli. Skull without sagittal crest; face more catlike; snout sharper. Old male with disagreeable odor (between asafetida and onion, according to Mohr, 1952*b*, p. 185). Pelage on back harsh to touch, hairs coarse, tips pointing directly backward; spots large and ring-shaped (almost absent in *Pusa sibirica*). Young born on ice in spring; white embryonal pelage persisting for about two weeks after birth.

Large males, length 140 cm. (55 in.); weight 90 kg. (199 lbs.); CBL 187 mm.; MW 115 mm. Large females, length 135 cm. (53 in.); weight *c.* 91 kg. (200 lbs.); CBL 167 mm.; MW 101 mm.............. *Pusa,* p. 95

13(11) (See plate 20, upper.) "The forelimbs and neighbouring parts of body dark, never any small dark spots on body. Condylobasal length of skull under 200 mm. Bony nasal septum just fails to reach rear edge of bony palate. The upper toothrow is curved, seen from below and from the side (in other words, curved in the horizontal and vertical planes)" (Ellerman and Morrison-Scott, 1951, p. 326). "Palatal length (measured from most anterior part of rostrum to midline at posterior edge of palate 86 mm. or less; posterior palatine foramina in, or posterior to, the maxillo-palatine suture" (Doutt, 1942, p. 80). Adult male dark brown with conspicuous yellowish-white band which "surrounds the neck extending forward to the middle of the head above; another broader yellowish-white band encircles the hinder portion of the body, from which a branch runs forward on each side to the shoulder, the two branches becoming confluent on the median line of the body below, but widely separated above . . . Adult female . . . uniform pale grayish-yellow or grayish-brown, with the exception of an obscure narrow transverse whitish band across the lower portion of the back. The extremities and the back are darker, with a faint indication of the dark 'saddle'-mark seen in the male" (Allen, 1880, pp. 676–77). Breeding on Pacific-Arctic ice.

Large males, length 168 cm. (66 in.) (Rass *et al.* 1955, p. 110, gave 190 cm.); weight 95 kg. (209 lbs.); CBL ?; MW ?. Large females, length 163 cm. (64 in.); weight 79 kg. (174 lbs.); CBL ?; MW ?. Sex unrecorded, CBL 194 mm.; MW 127 mm. . . . . *Histriophoca,* p. 102

13′ (See plate 21.) "The forelimbs and neighbouring parts of body are light-colored, and the body sometimes covered with small dark spots; condylobasal length of skull in adults over 200 mm. The bony nasal septum reaches the rear edge of the bony palate. Upper tooth-row not curved" (Ellerman and Morrison-Scott, 1951, p. 326). "Palatal length . . . more than 86 mm.; posterior palatine foramina in, or anterior to, maxillo-palatine suture" (Doutt, 1942, p. 80). Adults of both sexes with dark band, ∧-shaped, starting between shoulders and sending two branches hindward and along the sides; in the female the band is usually lighter and may be interrupted. Breeding on Atlantic-Arctic ice.

Large males, length *c.* 183 cm. (72 in.); weight *c.* 181 kg. (400 lbs.); CBL 229 mm.; MW 123 mm. Large females, length *c.* 183 cm. (72 in.); weight *c.* 181 kg. (400 lbs.); CBL 207 mm.; MW 117 mm. . . . . *Pagophilus,* p. 103

14(8) Lower incisors 2 on each side (1st lower lacking), as in Phocinae. Nasal passages not capable of great enlargement; anterior nares horizontal or nearly so, situated on dorsal surface of snout. Female slightly larger than male (verification needed for some species). Monogamous (promiscuous); resident or weakly migratory. . . . . Monachinae, 15

14′ Lower incisors 1 on each side (1st and 2d lower lacking). Nasal passages of adult male capable of great enlargement through combination of inflation and erection; anterior nares vertical or nearly so. Male larger than female. Polygynous or monogamous, depending on species; resident or strongly migratory, depending on species. . . . Cystophorinae, 19

15(14) (See plate 24.) Nasal processes of premaxillary (intermaxillary) broadly in contact with the nasals; postcanines wide and heavy, crushing type (shellfish and reef-fish diet?). Mammary teats 4. Embryonal (and newborn) pelage jet black; adult pelage never spotted; vibrissae smooth. Breeding only in subtropical waters.

Large males, length 290 cm. (114 in.); weight ?; CBL 279 mm.; MW 155 mm. Large females, length 278 cm. (109 in.); weight without viscera 302 kg. (666 lbs.); CBL ?; MW ?. Sex unrecorded, CBL 295 mm.; MW 178 mm. . . . . Monachini (*Monachus*), p. 112

15′ Nasal processes of premaxillary barely touching (*Leptonychotes*) or not touching the nasals; postcanines not crushing type. Mammary teats 2. Embryonal (and newborn) pelage white (grayish in *Leptonychotes*); adult pelage spotted in some species; vibrissae very faintly beaded. Breeding only in antarctic waters..........................Lobodontini, 16

16(15) Skull wider, adult CBL not over 50 percent greater than MW. Postcanines smaller, more homodont, not elaborately cusped.........17

16′ Skull narrower, adult CBL about 75 to 100 percent greater than MW. Postcanines larger, more heterodont, elaborately cusped. 18

17(16) (See plates 27 and 28.) Upper incisors conspicuously unequal, outer about 4 times larger than inner; postcanines strong and functional. Orbits not especially large, zygomatic arches not touching table when skull is at rest. IOW less than 18 percent of CBL. Low sagittal crest in both sexes. Embryonal (and newborn) pelage light rusty-gray. Gregarious at pupping time, habitat more continental (fast ice), circumpolar.

Large males, length 279 cm. (110 in.); weight 357 kg. (787 lbs.); CBL 290 mm.; MW 189 mm. Large females, length 293 cm. (115.5 in.); weight 426 kg. (940 lbs.); CBL 287 mm.; MW *c.* 180 mm....*Leptonychotes*, p. 122

17′ (See plate 26, upper.) Upper incisors nearly equal, outer slightly larger than inner; postcanines small and degenerate, occasionally absent in old animals (apparently adapted to diet of cephalopods). Orbits large (former name, the "big-eyed seal"), zygomatic arches dropping well below level of palate and supporting part of weight of skull on table. IOW greater than 18 percent of CBL. No sagittal crest; fontanelle persisting into early adulthood? Neck thick, short, skin folds may partly override the head. (Perkins, 1945, fig. 19.) Embryonal (and newborn) pelage white. Solitary; habitat more pelagic (pack ice); believed absent from eastern longitudes.

Large males, length 227 cm. (89.5 in); weight 181 kg. (400 lbs.); CBL 244 mm.; MW 172 mm. Large females, length 229 cm. (90 in.); weight 215 kg. (475 lbs.); CBL 242 mm.; MW *c.* 170 mm. Sex unrecorded, CBL 263 mm.; MW 176 mm. (Rudmose Brown, 1913, p. 197, gave "length of about 8 feet 6 inches" [how measured?].)..........*Ommatophoca*, p. 118

18(16) (See plate 26, middle and lower.) Body long and sinuous, up to 3.9 m.; skull massive, largest of all phocids except *Mirounga*; up to 416 mm. CBL. Postcanines sawlike, 3-cusped, the central cusp cylindrical, high, pointed, recurved. Prey largely penguins. Posterior edge of mandible gently rounded; definite sagittal crest in both sexes. Adult pelage the most heavily spotted of any antarctic seal; not conspicuously scarred.

Adult females about 10 percent larger than males. (Laws, 1957, fig. 2 and p. 54; sample of 34 individuals older than pups. Brown, 1957, p. 26.)

Large males, length 320 cm. (126 in.); weight 275 kg. (606 lbs.); CBL 416 mm.; MW 199 mm. Large females, length 381 cm. (150 in.); weight 453.5 kg. (1,000 lbs.); CBL 431 mm.; MW 205 mm. Sex unrecorded, 386 kg. (850 lbs.). Barrett-Hamilton (1902, p. 30) stated that Bruce measured an animal over 13 feet (396 cm.) in length, though it is not clear how the measurement was taken.................... *Hydrurga,* p. 120

18′ (See plate 25.) Body shorter and less flexible; length less than 3 m.; skull lighter and shorter, up to 306 mm. CBL. Postcanines 4- or 5-cusped, the principal cusp larger and often bulbous, pointed, recurved, with one cusp anterior and 2 or 3 posterior. "A set of teeth surmounted by perhaps the most complicated arrangement of cusps found in any living mammal" (Barrett-Hamilton, 1902, p. 13). Prey largely euphausians. Posterior edge of mandible abruptly rounded, nearly rectangular; no sagittal crest. Adult pelage lightly spotted (whitish and nearly immaculate just before shedding); frequently scarred.

Large males, length 257 cm. (101 in.); weight *c.* 224 kg. (494 lbs.); CBL 306 mm.; MW 167 mm. Large females, length 262 cm. (103 in.); weight *c.* 227 kg. (500 lbs.); CBL 286 mm.; MW 157 mm... *Lobodon,* p. 116

19(14) (See plate 29.) Size smaller, length up to 35 cm. and 408 kg. Outline of palatine bones nearly a square; tympanic bulla straight to convex in front; upper canine about twice the size of adjacent incisor. Nasal "hood" up to 25 cm. (9.8 in.) long (Freund, 1933, p. 80). In adult male "the redundant mucous membrane [can be] extruded as fiery red paired 'bladders' six to seven inches long and five to six inches in diameter" (Olds, 1950, p. 452). However, Erna Mohr has recently written (*in lit.*), upon dissection of a large male that died in Bremerhaven Zoo, ". . . there are not two bladders but only one!" (Negus, 1949, fig. 58, has sketched a longitudinal section of the snout in *Cystophora.*) Tip of snout haired. All of the digits with well-developed claws. Embryonal pelage white, shed before birth; adult mottled dark and light brown. Breeding on Atlantic-Arctic pack ice.

Large males, length *c.* 350 cm. (138 in.); weight *c.* 408 kg. (900 lbs.); CBL 275 mm.; MW 168 mm. Large females, length *c.* 300 cm. (118 in.); weight?; CBL 251 mm.; MW 165 mm................ *Cystophora,* p. 125

19′ (See plates 30 and 31.) Size largest of pinnipeds, length up to 650 cm. and over 3,000 kg. Outline of palatine bones "butterfly-shaped"; tympanic bulla concave in front; upper canine at least 5 times the size of adjacent incisor. "Trunk" or snout up to about 38 cm. (15 in.) long. Lining of snout cannot be voluntarily everted. Tip of snout hairless. Claws of pes rudimentary. Embryonal (newborn) pelage brownish black; adult almost uniformly colored, never spotted. Breeding on land in circumpolar subantarctic seas (one small population in California-Mexico).

Large males, length 650 cm. (256 in.); weight *c.* 3,629 kg. (8,000 lbs.); CBL 561 mm.; MW 293 mm. Large females, length 351 cm. (138 in.); weight *c.* 907 kg. (2,000 lbs.); CBL 333 mm.; MW 202 mm. Sex unrecorded, CBL 519 mm. with MW 309. . . . . *Mirounga,* p. 128

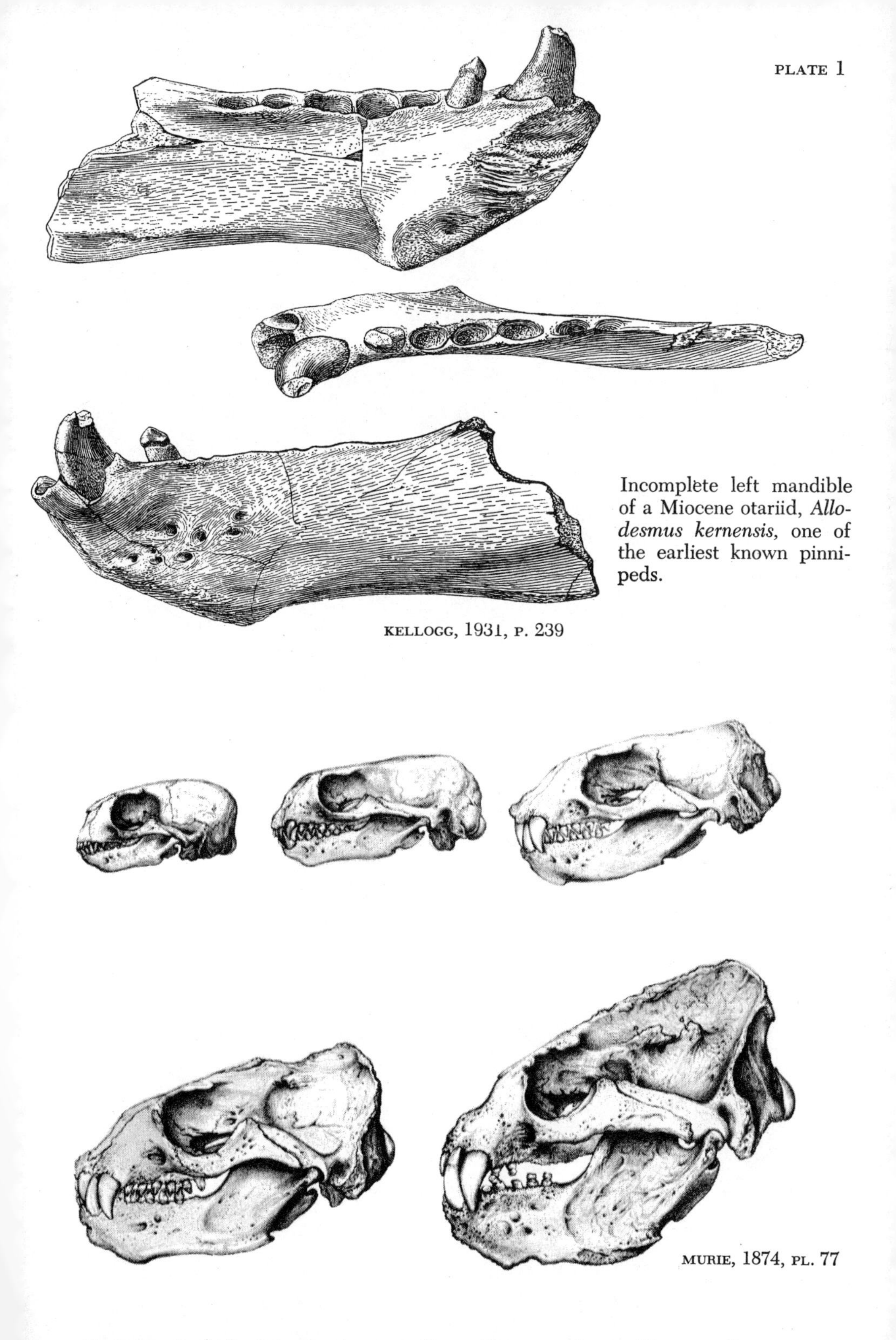

Incomplete left mandible of a Miocene otariid, *Allodesmus kernensis*, one of the earliest known pinnipeds.

KELLOGG, 1931, P. 239

MURIE, 1874, PL. 77

Variation in skull of the southern sea lion with age and sex. From smallest to largest: a pup about a fortnight old, sex unrecorded; a yearling, sex unrecorded; a female, age unrecorded; an adult but not full-grown male; an old male.

Above. Adult male Steller sea lion, estimated weight over 900 kg. (1 ton), defending his breeding territory on the Pribilof Islands, Alaska, 2 July 1949.

Below. Steller sea lions massed on Amak Island, Bristol Bay, Alaska, 30 July 1952, about one month after the start of the pupping season.

FISHERIES RESEARCH INSTITUTE PHOTO BY W. F. THOMPSON

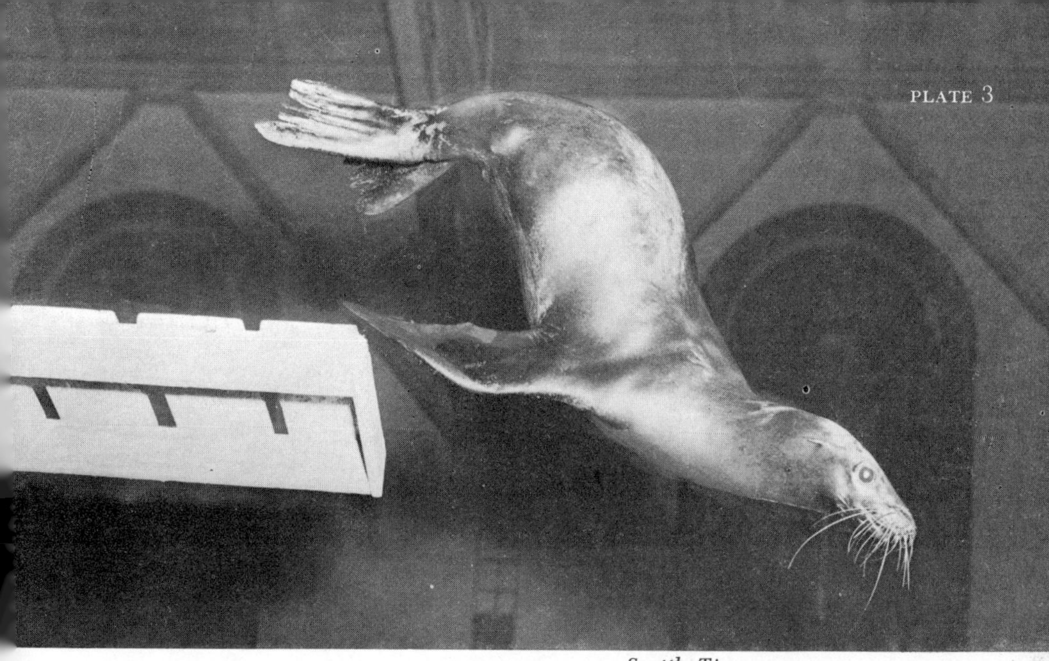

Seattle Times PHOTO BY ROY E. SCULLY

Above. California sea lion leaping from a circus platform.

Below. California sea lion, a male "trained seal" in its 12th year of captivity, weight about 110 kg. (243 lbs.), 19 March 1949.

PLATE 4

J. H. SORENSEN

Above. New Zealand sea lions on Campbell Island, 22 November 1942, adults and immature males.

Below. New Zealand sea lion, adult male on Campbell Island, 22 February 1946.

J. H. SORENSEN

South American fur seal, Signy Island, South Orkney Islands, April, 1954.

FALKLAND ISLANDS DEPENDENCIES SURVEY

South American fur seal, South Georgia.

FALKLAND ISLANDS DEPENDENCIES SURVEY

South American fur seal, adult male on Isla Genovesa, Galapagos Islands.

ALLAN HANCOCK FOUNDATION

NORTHERN FUR SEALS AT THE CLIMAX OF THE BREEDING SEASON

Above. St. Paul Island, Alaska, 16 July 1948. In the middle distance are nonbreeding animals, including mainly subadults and cripples.

Below. Sea at bottom. Center, clusters of seals in "harems," and an elevated observation walk (over-all length 66 m.). At left of walk, a dominant male or "harem bull" has cleared a circular path around a harem. The periphery of the path is dotted with small black pups. Above, individual subordinate males or "idle bulls," with gulls talking flight. St. Paul Island, Alaska, 15 July 1948.

AUTHOR AND K. W. KENYON

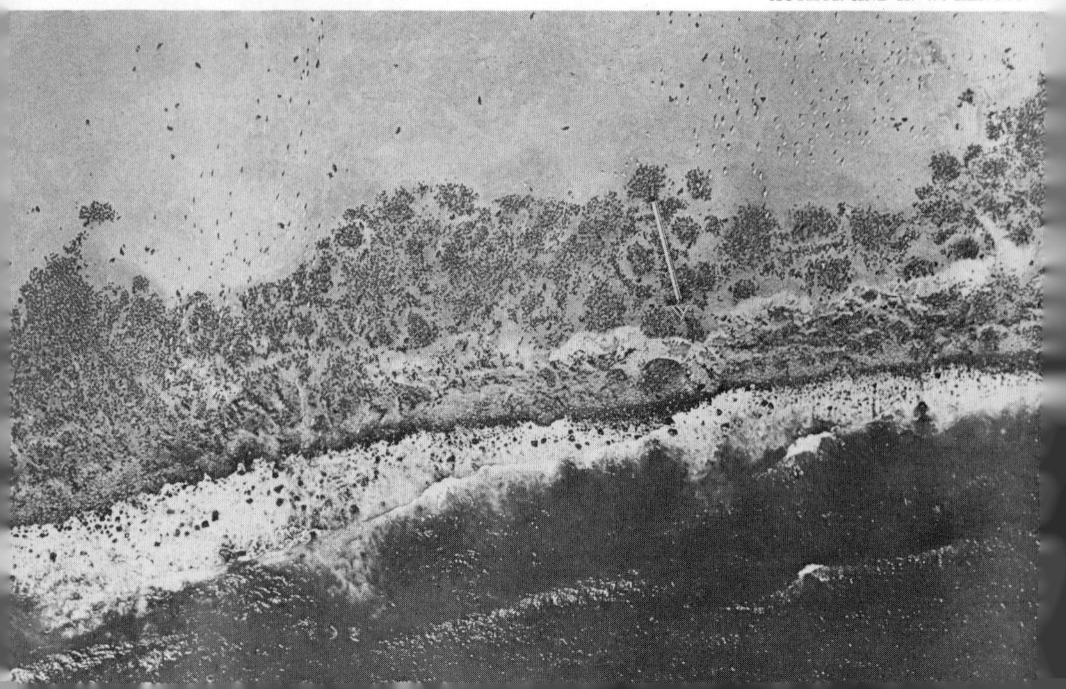

Above. Northern fur seals in copulation, St. Paul Island, Alaska, 15 July 1949. All otariid seals mate on land.

Below. Northern fur seal giving birth to young about mid-July.

O. W. OLSEN

PLATE 8

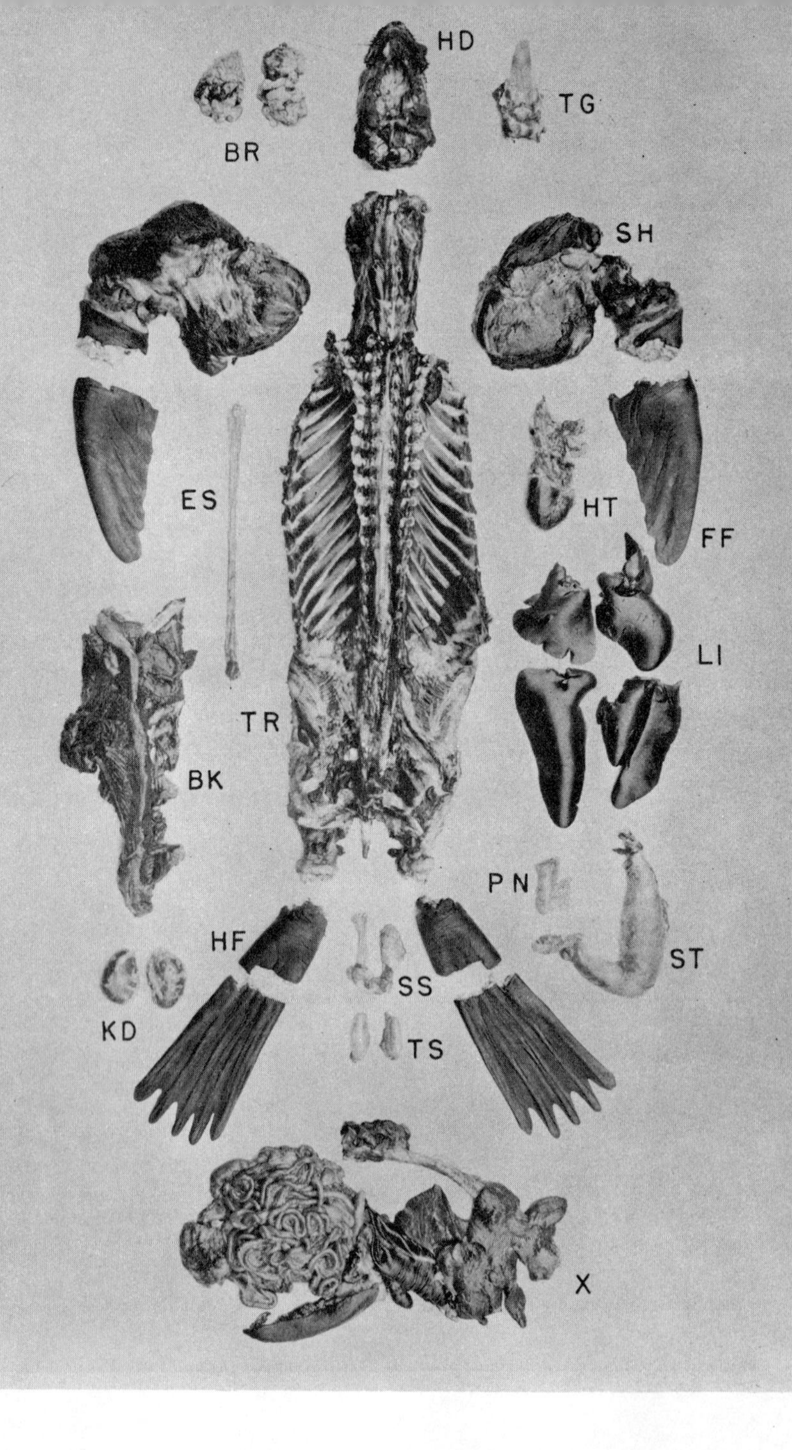

Dissection of the skinned carcass of a subadult male northern fur seal showing parts formerly used in the Pribilof Island native economy (see Scheffer, 1948). BK = brisket, BR = brain, ES = esophagus, FF = fore flipper, HD = head, HF = hind flipper, HT = heart, KD = kidneys, LI = liver, PN = pancreas, SH = shoulder, SS = "seal stick" (penis, bladder, and prostate), ST = stomach, TG = tongue, TR = trunk, TS = testes, X = intestines, gall bladder, spleen, bronchus, and lungs.

U.S. NATIONAL MUSEUM

Above. Skeleton of a northern fur seal, adult male.

Below. Annuli, or growth ridges, on the root of the right upper canine tooth of a 6-year-old male northern fur seal, 17 June 1947. Numbers identify the midsummer fasting periods. CEJ = cemento-enamel junction. About 3.4 × nat. size.

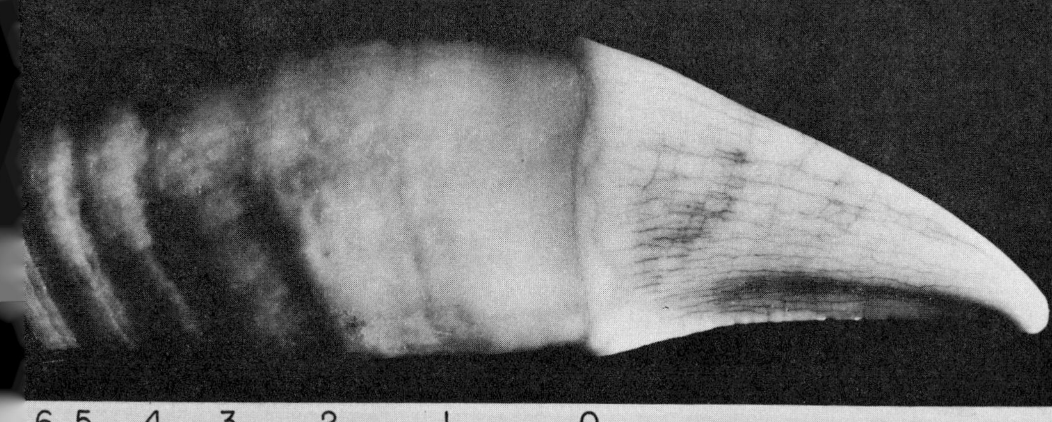

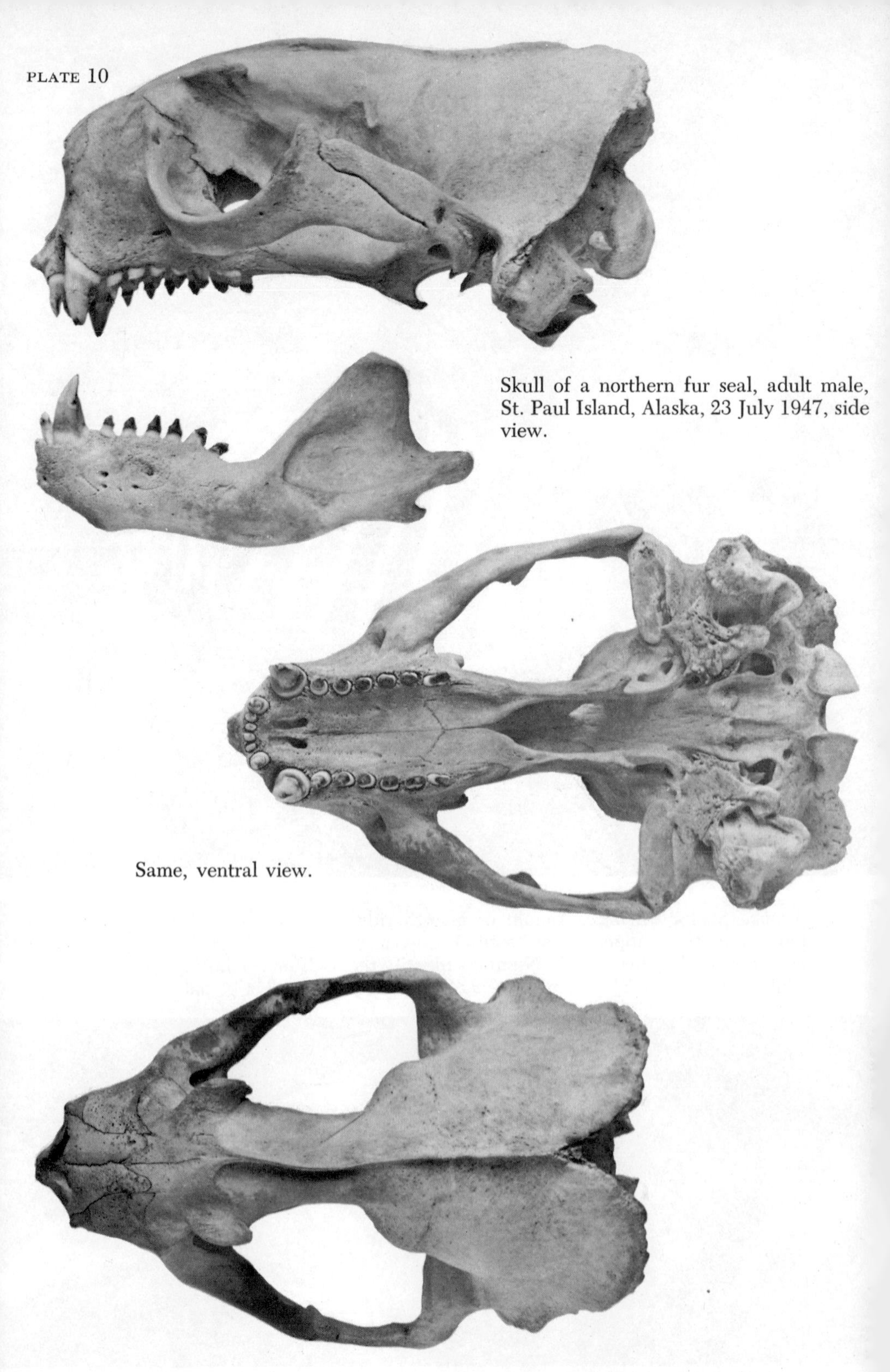

Skull of a northern fur seal, adult male, St. Paul Island, Alaska, 23 July 1947, side view.

Same, ventral view.

Same, dorsal view (see also fig. 15).

FORD WILKE

Above. A piebald northern fur seal pup; one among tens of thousands of all-black young; 29 July 1953.

Below. Section of the dried pelt of a northern fur seal showing (bottom to top): white skin, light brown fur, and white-tipped guard hairs. About 7.1 × nat. size.

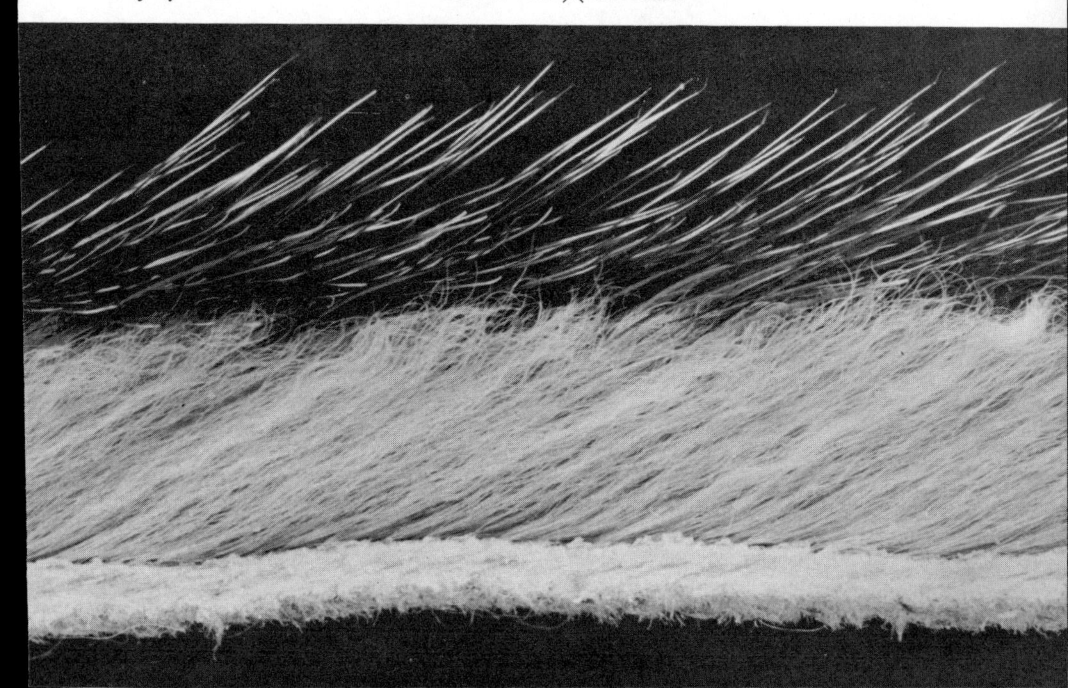

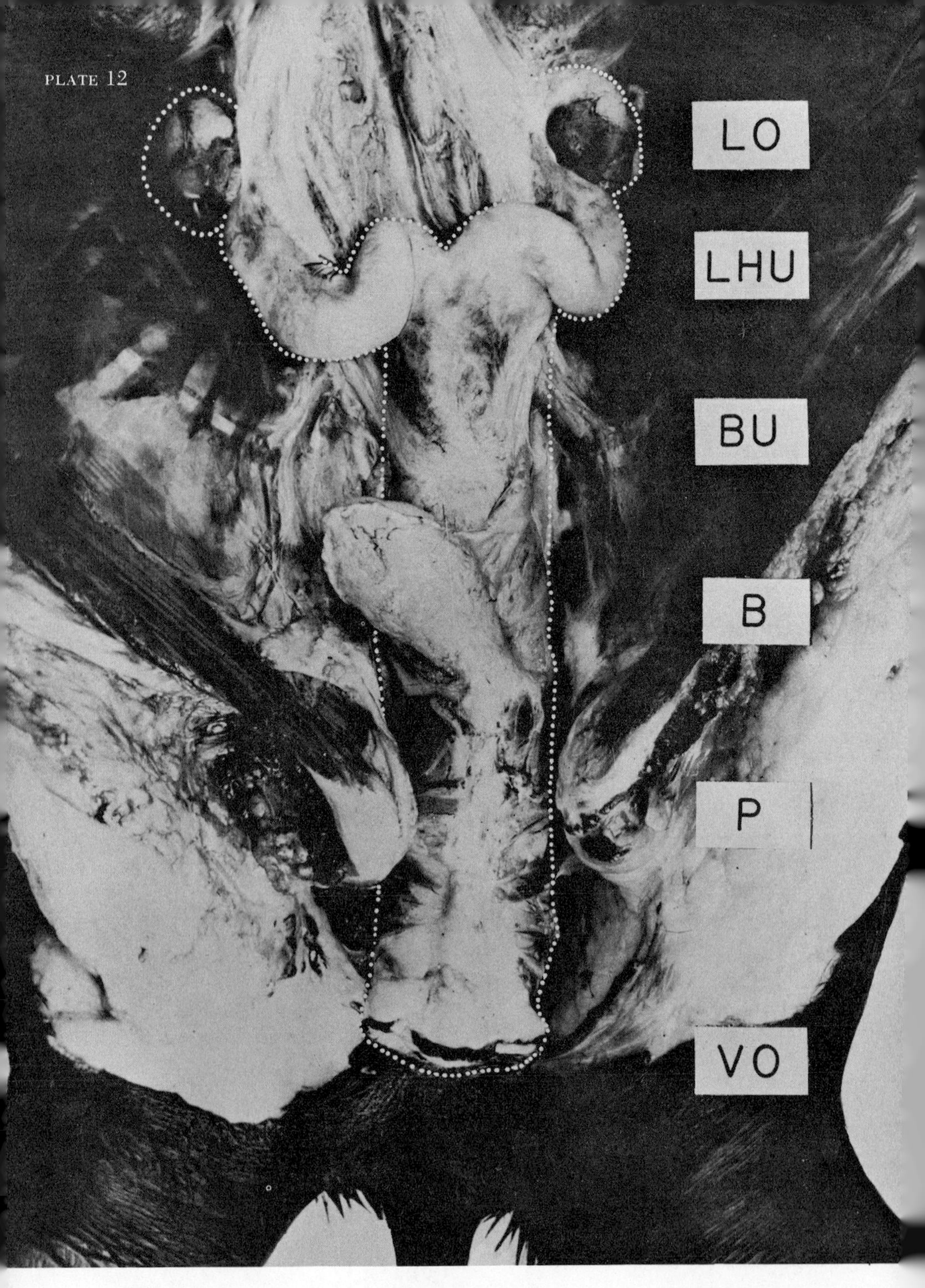

Reproductive tract of a female northern fur seal (probably a nulliparous 4-year-old), 4 August 1945, ventral view. LO = left ovary, LHU = left horn of uterus, BU = body of uterus, B = bladder, P = pubic bone, VO = vaginal opening, hiding anus.

Above. Bodies of northern fur seal pups, many infested with hookworms, on St. Paul Island, Alaska, 11 August 1948. Biologists counted 21,600 dead pups along 1.5 miles of seacoast on this date.

Below. Northern fur seal females afflicted with a skin disorder known locally as "mange," though of unknown origin; St. Paul Island, Alaska, 23 July 1940.

Skull of an adult male walrus from St. Paul Island, Alaska. Weight, including tusks and lower jaws, 25.2 kg. (55.5 lbs.), CBL 352 mm.

Mounted head of an adult male walrus from eastern Siberia.

NEW YORK ZOOLOGICAL SOCIETY PHOTO BY GRANCEL FITZ

ZOOLOGICAL SOCIETY OF SAN DIEGO

Above. Male walrus calf, age 7 months, in the San Diego, California, Zoo; born about May on the coast of Greenland.

Below. Adult female walrus, age 12 years, in the Copenhagen Zoo, January 1949.

AXEL REVENTLOW

Harbor seal, a 4-year-old male, in a salt-water aquarium, Tacoma, Washington, 7 May 1942.

KELSHAW BONHAM

Above. Full-term fetus of a harbor seal; female, length 91 cm. (35.8 in.), weight 12.5 kg. (27.5 lbs.), 31 May 1942. Note loose remnants of fetal wool plastered to the body, and the tiny ear pinna, lost in adult phocids.

Below. A harbor seal 1 to 2 weeks old, Copalis Beach, Washington, 30 May 1942. Note the mystacial and superciliary vibrissae.

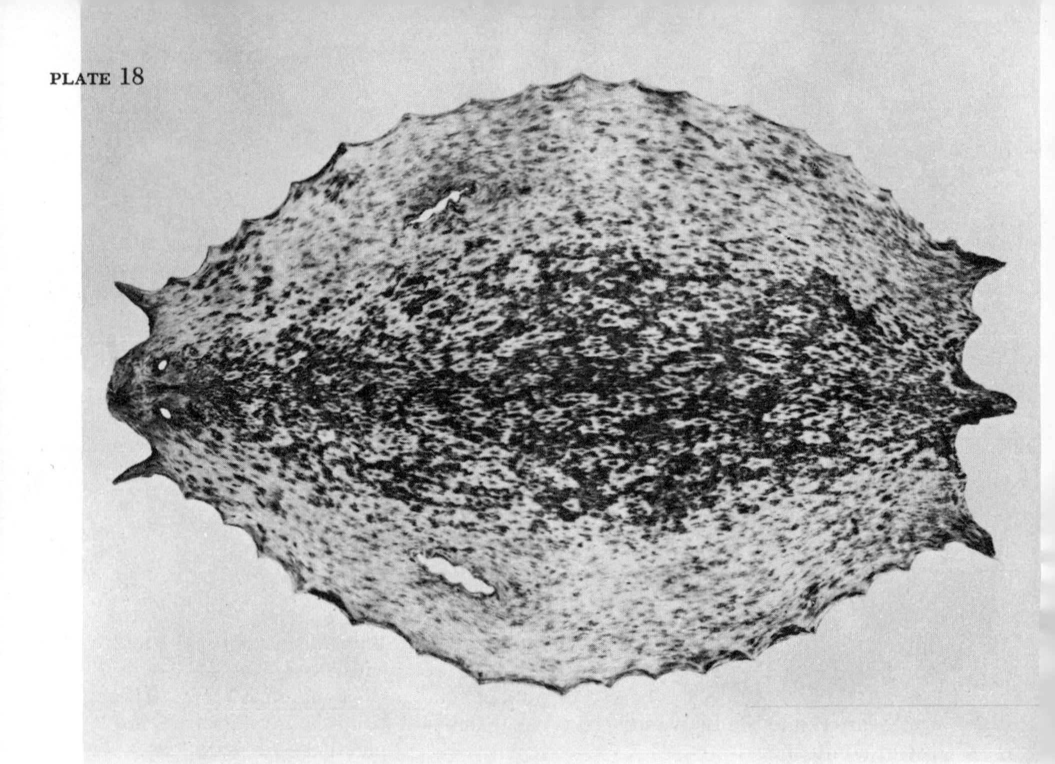

Above. Pelt of an adult female harbor seal showing characteristic spotting; Neah Bay, Washington, 19 May 1942. The fleshed, dried pelt without flippers weighs 2.1 kg. (4.6 lbs.) and is 182 cm. (71.8 in.) from nose to tip of tail.

Below. Pelage of a female harbor seal about 3 months old, 10 September 1941. Under side of leather of a chemically tanned pelt showing hairs extending backward beyond the cut edge. Note sparse undercoat of fine curly hairs. About 6.2 × nat. size.

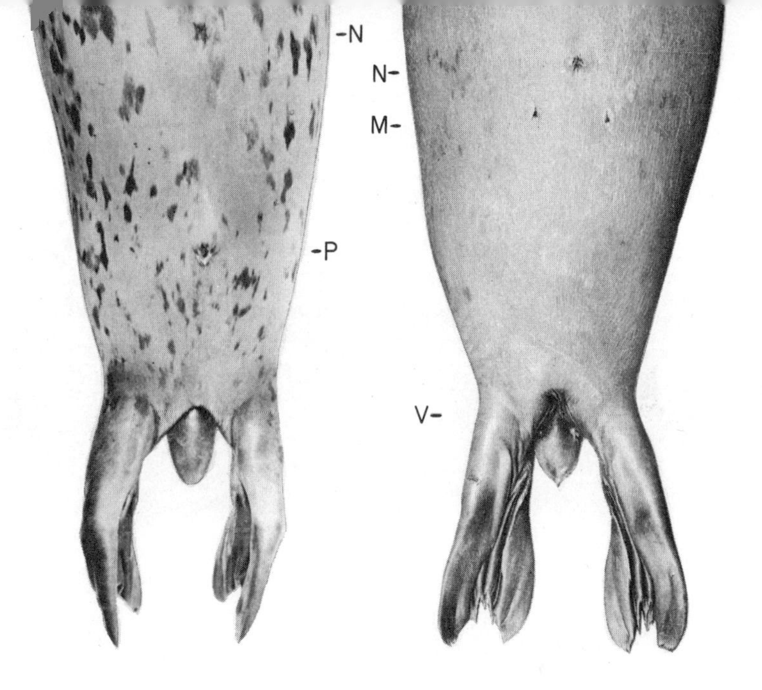

Above. External genitalia of the harbor seal. *Left*, subadult male, showing navel and penile opening. *Right*, subadult female, showing navel, mammary teats, and vaginal opening.

Below. King salmon mutilated by harbor seals in Skagit Bay, Washington, 17 June 1942. These remnants were found hanging in a gill net.

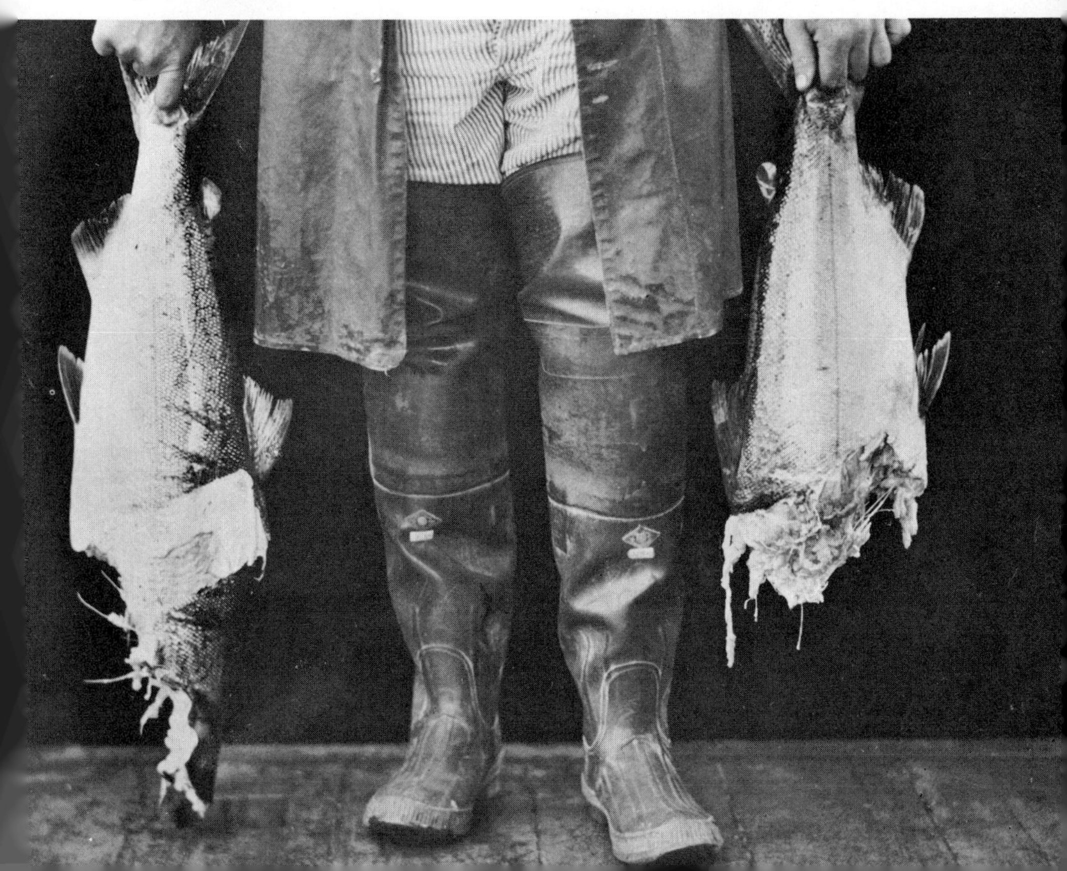

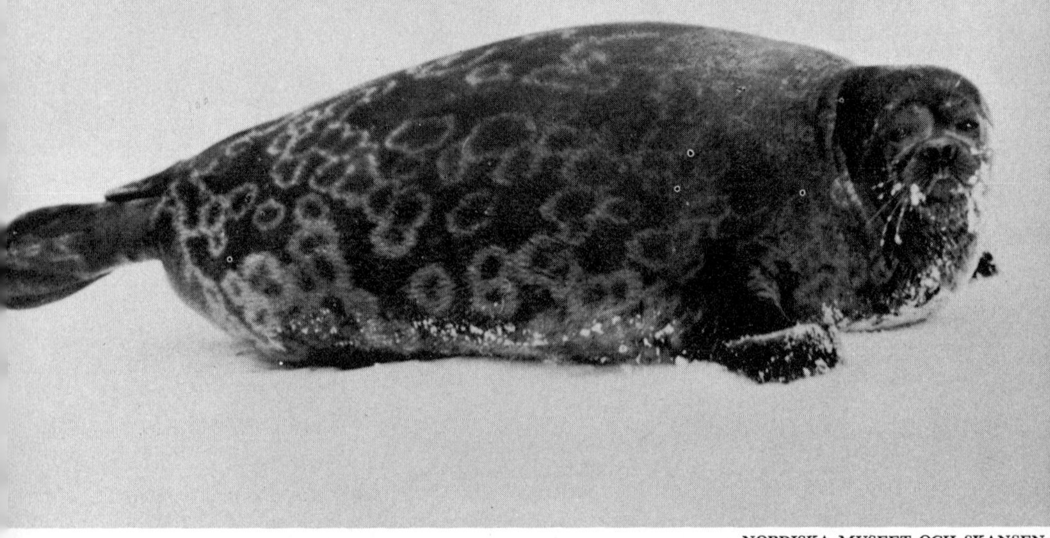

NORDISKA MUSEET OCH SKANSEN

Above. Ringed seal, Baltic Sea.

Below. Mounted pelt of a ribbon seal, St. Lawrence Island, Alaska, January 1949; head at left, flippers and mask removed.

SIVERTSEN, 1941, PL. 7

Above. Harp seals, adult male (*right*) and female, White Sea, May 1933.

Below. Adult male harp seal.

ALWIN PEDERSEN

SETON GORDON

Above. Grey seals approaching copulatory position, male on right; Isle of Skye, Scotland.

Below. Newborn grey seal on Eilean nan Ron, off the coast of Oronsay, Scotland, in early October.

TOM WEIR

SKETCH FROM LIFE BY FRANCIS H. FAY

Above. Bearded seal. *Below left*. Bearded seal in Zoologischen Gartens Berlin; whiskers dry and curled at their tips. *Below right*. Bearded seal with whiskers wet, smooth, and shiny.

ERNA MOHR

ERNA MOHR

Above. Face of a male Hawaiian monk seal about 4–5 months old; eviscerated weight 50 kg. (110 lbs.); a zoo specimen; 30 August 1951.

Below. Female Hawaiian monk seal with pup, Kure Atoll, 5 June 1957.

KARL W. KENYON

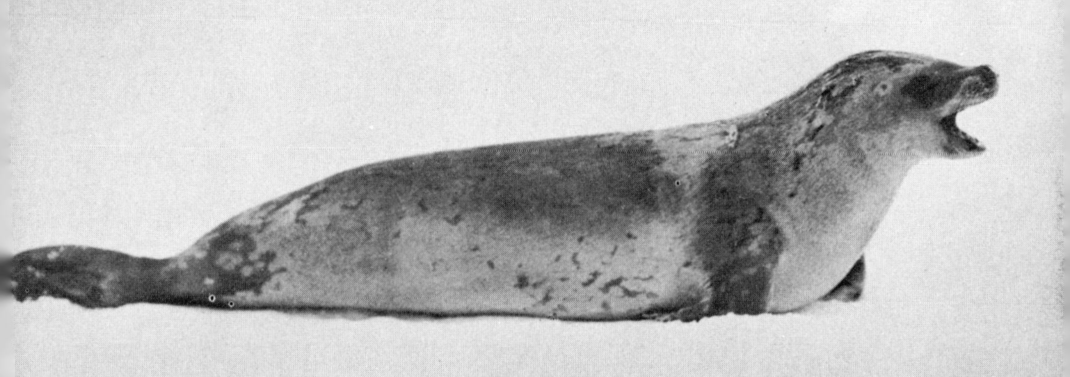

U.S. ANTARCTIC SERVICE, BYRD EXPEDITION 1940–41

Above. Crabeater seal, West Base, Antarctica.

Below. Skull of a young crabeater seal showing elaborate postcanine teeth adapted for straining macroplankton from the sea.

LINDSEY, 1938, FIG. 2

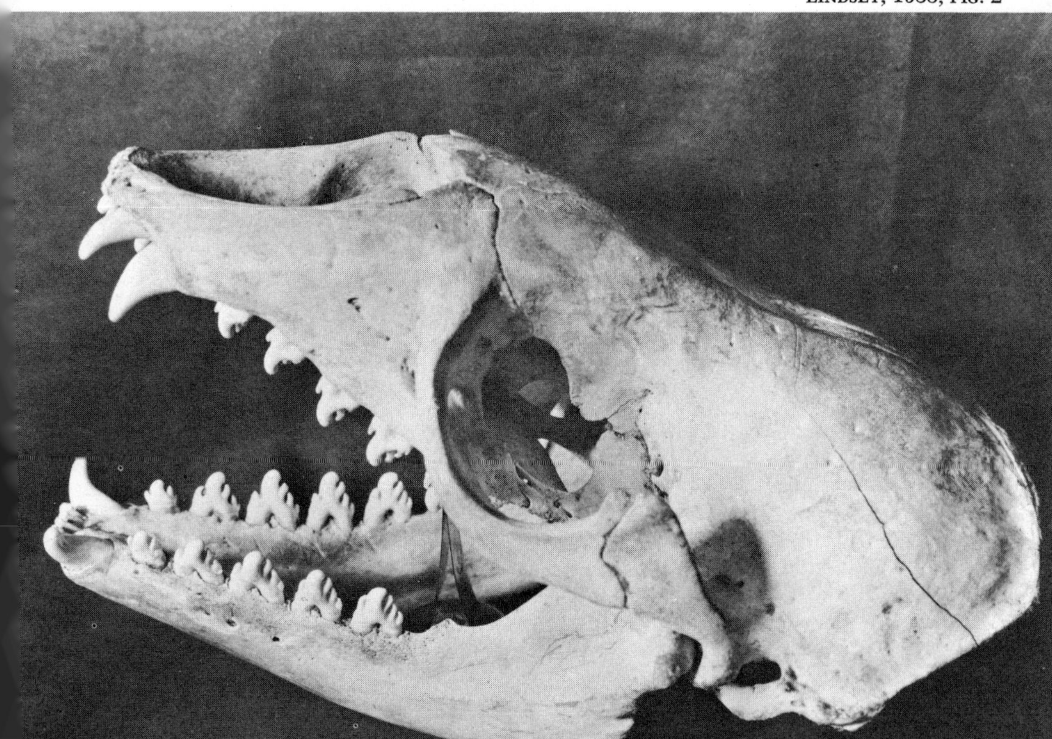

U.S. ANTARCTIC SERVICE, BYRD EXPEDITION 1940–41

Body of a Ross seal, rarely seen or photographed, Antarctica, 10 January 1941.

U.S. ANTARCTIC SERVICE, BYRD EXPEDITION 1940–41

Above. Leopard seals on an ice floe in Anderson Harbor, Palmer Archipelago, Antarctica.

Below. A leopard seal, showing dentition adapted for seizing and tearing the flesh of penguins and other large, warm-blooded prey.

FALKLAND ISLANDS DEPENDENCIES SURVEY PHOTO BY F. G. BIRD

U.S. COAST GUARD

Above. A Weddell seal awakening, Antarctica, 1947.

Below. Weddell seals on shore ice in the Bay of Whales, Antarctica, in early summer, 15 January 1947.

U.S. NAVY

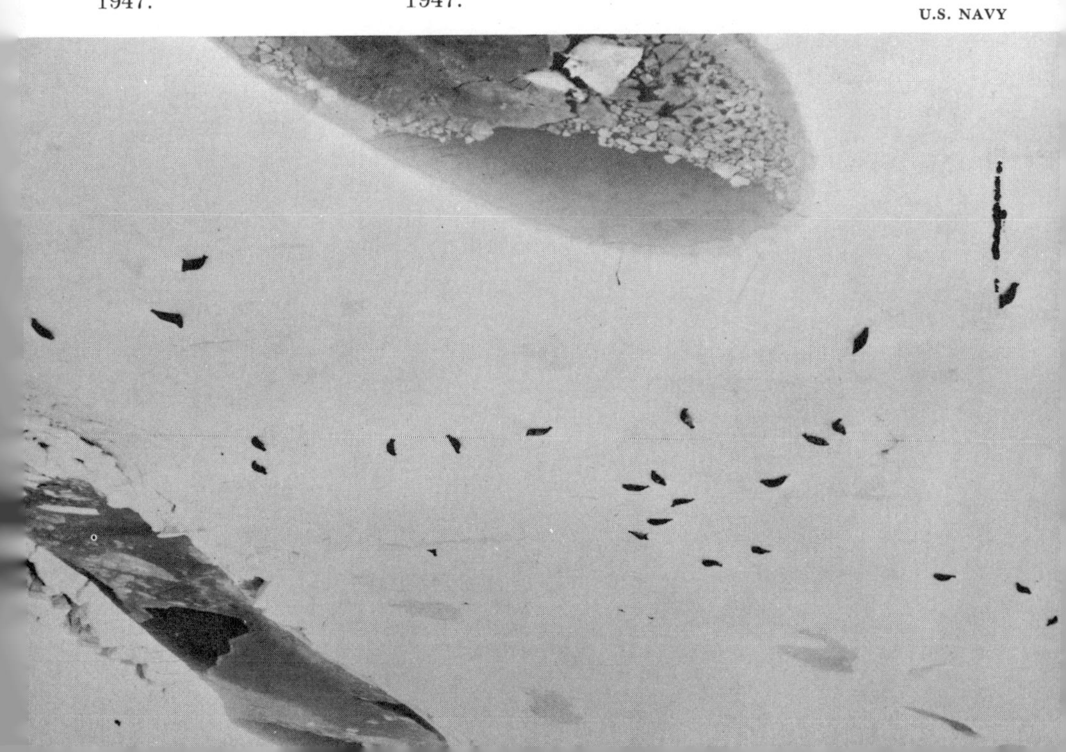

U.S. ANTARCTIC SERVICE, BYRD EXPEDITION 1940–41

Above. A Weddell seal returning to the sea through a "breathing hole" in the ice, 25 October 1940.

Below. A female Weddell seal with suckling pup, 24 October 1940.

U.S. ANTARCTIC SERVICE, BYRD EXPEDITION 1940–41

WERNER TRENSE

Above. A hooded seal in anger everts its nasal sac, Jan Mayen, March 1950.

Below. Female hooded seal with young; the latter in "blueback" stage, North Harpswell, Maine, 25 March 1928.

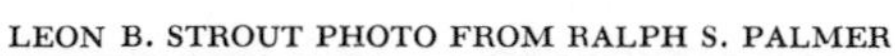

LEON B. STROUT PHOTO FROM RALPH S. PALMER

U.S. COAST GUARD PHOTO BY MASHBURN

Southern elephant seal, adult but not very large male, Antipodes Islands, 4 March 1947.

U.S. COAST GUARD PHOTO BY MASHBURN

ZOOLOGICAL SOCIETY OF LONDON

Above. Young southern elephant seal from the Falkland Islands, showing flexibility of the vertebral column.

Below. Face of an old northern elephant seal, Isla de Guadalupe, Mexico.

ALLAN HANCOCK FOUNDATION PHOTO BY W. CHAS. SWETT

AERO-MARINE PHOTOS

Above. A group of killer whales (*Orcinus*) cruising in Puget Sound, 13 August 1947. The killer whale is an important predator of seals in all oceans.

San Francisco Examiner

Above. A killer whale attacked and sank this dory in Bodega Bay, California, 28 March 1952, forcing two fishermen to swim to safety. Note tooth marks in the hull.

Below. Jaws of a great white shark (*Carcharodon*); length 4.27 m. (14 ft.), weight about 800 kg. (1,764 lbs.); caught in Willapa Harbor, Washington, 1 September 1950. Its stomach contained, besides fishes and invertebrates, the skins of two harbor seals.

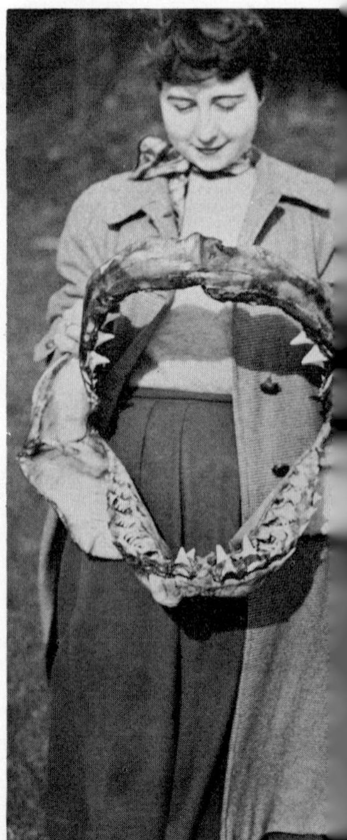

LITERATURE CITED

(Citations follow approximately the style described in *Bulletin of the American Institute of Biological Sciences,* vol. 2, p. 21–23, 1952. Titles of most serial publications are abbreviated in the style of *World List of Scientific Periodicals.* Ed. 3. London, Butterworth, 18 + 1058 p., 1952.)

Admiralty, Hydrographic Department. 1948. The antarctic pilot. Ed. 2. London, 42 + 370 p., 3 fold. maps.

Ainsworth, G. F. 1934. Life on Macquarie Island [p. 334–356]. *In* The home of the blizzard, by D. Mawson. Abridged pop. ed. London, Hodder and Stoughton.

Alaska Department of Fisheries. 1956? Annual report for 1955. Juneau, Rept. 7, 152 p.

Allen, G. M. 1938. Order Pinnipedia [vol. 11, part 1, p. 490–493]. *In* The mammals of China and Mongolia, by G. M. Allen, 2 parts, 1938, 1940. Amer. Mus. Nat. Hist., Nat. Hist. of Central Asia, 11 vols., 1932–40.

Allen, G. M. 1939. A checklist of African mammals. Bull. Mus. Comp. Zool. Harv. 83:1–763.

Allen, G. M. 1942. Extinct and vanishing mammals of the western hemisphere with the marine species of all the oceans. New York, Publ. Amer. Committee Internatl. Wild Life Protection 11, 16 + 620 p.

Allen, J. A. 1870. On the eared seals (Otariadae), with detailed descriptions of the North Pacific species . . . Bull. Mus. Comp. Zool. Harv. 2:1–108, 3 pls.

Allen, J. A. 1880. History of North American pinnipeds, a monograph of the walruses, sea-lions, sea-bears and seals of North America. Washington, U.S. Geol. and Geogr. Surv. Terr., Misc. Publ. 12, 16 + 785 p.

Allen, J. A. 1892? A synopsis of the pinnipeds, or seals and walruses, in relation to their commercial history and products [p. 367–391 and frontis.]. *In* Fur seal arbitration, appendix to the case of the United States before the tribunal of arbitration to convene at Paris . . . Washington, Govt. Print. Off., vol. 1 (title p. dated 1895, appendix dated 1892).

Allen, J. A. 1902*a*. The generic and specific names of some of the Otariidae. Bull. Amer. Mus. Nat. Hist. 16:111–18.

Allen, J. A. 1902*b*. The hair seals (family Phocidae) of the North Pacific Ocean and Bering Sea. Bull. Amer. Mus. Nat. Hist. 16:459–499.

Allen, J. A. 1903. Report on the mammals collected in northeastern Siberia by the Jesup North Pacific expedition, with itinerary and field notes by N. G. Buxton. Bull. Amer. Mus. Nat. Hist. 19:101–184.

Allen, J. A. 1905. The Mammalia of southern Patagonia. Repts. Princeton Univ. Exped. Patagonia, 1896–1899, vol. 3 (zool.), no. 1, 210 p., 29 pls.

Amoroso, E. C., and J. H. Matthews. 1951. The growth of the grey seal (*Halichoerus grypus*) from birth to weaning. J. Anat. 85:426–428.

Anderson, R. M. 1943. Two new seals from arctic Canada with key to the Canadian forms of hair seals (family Phocidae). Quebec, Ann. Rept. Provancher Soc. Nat. Hist. 1942:23–34 (English), 35–47 (French).

Anderson, R. M. 1946 [1947]. Catalogue of Canadian Recent mammals. Bull. Nat. Mus. Canada 102 (biol. ser. 31), 6 + 238 p.

Andersson, K. A. 1908. Das höhere Tierleben im antarktischen Gebiete. Wissensch. Ergeb. der Schwedische Südpolar-Exped. 1901–1903. Stockholm, Band 5 (zool.), Lief. 2:1–58, 10 pls., 2 maps.

Angot, M. 1954. Observations sur les mammifères marins de l'archipel de Kerguelen . . . Mammalia, 18:1–111, 6 pls., 2 maps.

Anonymous. 1946*a*. Hi-jumping sea lion betters highest mark set by human athlete. See 9 March, p. 30–31.

Anonymous. 1946*b*. [*Lobodon carcinophagus* in Tasmania]. Pap. Roy. Soc. Tasm. 1945:165.

Anonymous. 1948. Sea lion slaughter. Nature Mag. 41:5.

Anonymous. 1949. Ashore at Cape Town during a storm . . . Illus. Lond. News, 214(5746):760.

Anonymous. 1951. The polar bear. Nat. Mus. Canada, Guide to Mammalian Habitat Exhibits, Leafl. 4, 4 p.

Anonymous. 1957. Ostrov Tyuleniy. Geografiya v Shkole [Geography in School], no. 3, p. 63. (Reference from Scott Polar Research Institute.)

Anson, G. 1748. A voyage round the world in three years [1740–44]. Compiled . . . by R. Walter. London, John and Paul Knapton, 34 + 417 p.

Antarctic Pilot (see Admiralty, 1948).

Asdell, S. A. 1946. Patterns of mammalian reproduction. Ithaca, Comstock, 14 + 437 p.

Atkinson, A. L. C., and W. A. Bryan. 1913. A rare seal [*Monachus schauinslandi*]. Bull. N.Y. Zool. Soc. 16:1050–1051.

Backer, O. F. 1948. Seal hunting off Jan Mayen. Nat. Geogr. Mag. 93:57–72, 15 photos (10 in color), 1 map.

Backhouse, K. M. 1954. The grey seal. Univ. Durham Coll. Med. Gaz. 48(2): 9–16.

Backhouse, K., and H. R. Hewer. 1956. Delayed implantation in the grey seal, *Halichoerus grypus* (Fab.). Nature 178:550.

Bailey, A. M., and R. W. Hendee. 1926. Notes on the mammals of northwestern Alaska. J. Mammal. 7:9–28, 3 pls.

Bailey, B. E. 1952. Marine oils with particular reference to those of Canada. Bull. Fish. Res. Bd. Can. 89, 413 p.

Balkwill, F. H. 1888. On the geographical distribution of seals. Zoologist, ser. 3, 12:401–411.

Banning, G. H. 1933. Hancock expedition of 1933 to the Galápagos Islands. General report. Bull. Zool. Soc. San Diego, Calif. 10:1–30, frontis., map.

Barabash-Nikiforov, I. I. 1938. Mammals of the Commander Islands and the surrounding sea. J. Mammal. 19:423–429.

Barrett-Hamilton, G. E. H. 1902. Report on the collections of natural history made in the antarctic regions during the voyage of the "Southern Cross." London, Brit. Mus. (Nat. Hist.) 66 p., 1 pl.

Barrow, K. M. (Mrs.) 1910. Three years in Tristan da Cunha. London, Skeffington, 12 + 280 p., map, illus.

Bartholomew, G. A. 1950. A male Guadalupe fur seal on San Nicolas Island, California. J. Mammal. 31:175–180.

Bartholomew, G. A. 1952. Reproductive and social behavior of the northern elephant seal. Univ. Calif. Publ. Zool. 47:369–472, 20 pls.

Bartholomew, G. A. 1955. The northern elephant seal. Zoonooz, San Diego, Calif. 28:6–9.

Bartholomew, G. A., and C. L. Hubbs. 1952. Winter population of pinnipeds about Guadalupe, San Benito, and Cedros Islands, Baja California. J. Mammal. 33:160–171.

Bartholomew, G. A., and F. Wilke. 1956. Body temperature in the northern fur seal, *Callorhinus ursinus.* J. Mammal. 37:327–337.

Bartholomew, J. G., W. E. Clarke, and P. H. Grimshaw. 1911. Pinnipedia [p. 15–16, pl. 4, maps 5–6]. *In* Atlas of zoogeography, Bartholomew's physical atlas, vol. 5. Edinburgh, Roy. Geogr. Soc.

Baur, G. 1897. [Distribution of marine mammals]. Science, n.s., 5:956–957.

Beddard, F. E. 1902. Pinnipedia [vol. 10, p. 446–455]. *In* The Cambridge natural history. London, Macmillan, 10 vols., 1895–1909.

Beebe, C. W. 1926. The Arcturus adventure. New York, Putnam's Sons, 20 + 439 p., many pls.

Beneden, M. P.-J. van. 1876. Les phoques fossiles du bassin d'Anvers. Bull. Acad. Roy. Belg. 41:783–812, 1 pl.

Berdegué, J. 1956. La foca fina, el elefante marino y la ballena gris en Baja California, y el problema de su conservacion. Ediciones del Instituto Mexicano de Recursos Naturales Renovables, A.C., no. 14, 38 p.

Berg, C. 1898. *Lobodon carcinophagus* (H. J.) Gr. en el Rio de la Plata. Comun. Mus. Nac. B. Aires, Tomo 1, p. 15. (Reprinted in full in Allen, 1905, p. 93.)

Bergersen, B. 1931. Beiträge zur Kenntnis der Haut einiger Pinnipedier . . . Skrift. Norske Vidensk.-Akad., Oslo, Mat.-Naturvidensk. Klasse, 1931:1–179, 22 pls.

Berry, E. W., and W. K. Gregory. 1906. *Prorosmarus alleni,* a new genus and species of walrus from the upper Miocene of Yorktown, Virginia. Amer. J. Sci., ser. 4, 21:444–450.

Bertram, G. C. L. 1940. The biology of the Weddell and crabeater seals, with a study of the comparative behaviour of the Pinnipedia. Brit. Mus. (Nat. Hist.) Sci. Repts. Brit. Graham Land Exped. 1934–1937, 1:1–139, 10 pls., 1 fold. tbl.

Bini, G. 1951. Osservazioni su alcuni mammiferi marini sulle coste del Cile e del Perù. Bol. Pesca Piscic. Idrobiol., Roma, anno 27, vol. 6(n.s.), fasc. 1, p. 79–93.

Bishop, D. W. 1950. Respiration and metabolism [chap. 8, p. 209–289]. *In* Comparative animal physiology, C. L. Prosser, ed. Philadelphia, W. B. Saunders.

Black, M. M., W. S. Rapson, H. M. Schwartz, and N. J. Van Rensburg. 1945. South African fish products. Part 19. The South African seal fishery. J. Soc. Chem. Ind., London, 64:326–331.

Blainville, H.-M. D. de. 1820. Sur quelques crânes de phoques. J. de Physique, de Chimie, d'Histoire Naturelle et des Arts, 91:286–300.

Bobrinskoi, N. A. 1944. Pinnipedia [p. 162–168]. *In* Mammals of U.S.S.R. Moscow, 439 p. (Library Congress cat. card 50-48933, English transl. in Oxford Bur. Animal Population.)

Boddaert, P. 1785. Quadrupedia [vol. 1, p. 38 + 174]. *In* Elenchus animalium, Rotterdam.

Boetticher, H. von. 1934. Die geographische Verbreitung der Robben (Pinnipedia). Z. Säugetierk. 9:359–368.

Bonnot, P. 1929. Report on the seals and sea lions of California, 1928. California Div. Fish and Game, Fish. Bull. 14, 62 p.

Bonnot, P. 1951. The sea lions, seals and sea otter of the California coast. Calif. Fish Game 37:371–389.

Bonnot, P., G. H. Clark, and S. R. Hatton. 1938. California sea lion census for 1938. Calif. Fish Game 24:415–419.

Brandenburg, F. G. 1938. Notes on the Patagonian sea lion. J. Mammal. 19: 44–47.

Brass, E. 1911. Aus dem Reiche der Pelze . . . Berlin, im Verlage der Neuen Pelzwaren-Zeitung, 21 + 709 p.

Brazenor, C. W. 1950. The mammals of Victoria. Nat. Mus. Victoria, Handbook 1. Melbourne, Brown, Prior and Anderson, 125 p.

Brimley, H. H. 1931. Harbor seal in North Carolina. J. Mammal. 12:314.

Brisson, M. J. 1762. Regnum animale in classes IX distributum . . . Editio altera auctior. Ed. 2. Lugduni Batavorum, 8 + 296 p.

Brookes, J. 1828. A catalogue of the anatomical & zoological museum of Joshua Brookes . . . which will be sold at auction . . . on Monday, the 14th day of July, 1828 . . . London, 76 p.

Brooks, J. W. 1953. The Pacific walrus and its importance to the Eskimo economy. Trans. 18th N. Amer Wildl. Conf., p. 503–510.

Brooks, J. W. 1954. A contribution to the life history and ecology of the Pacific walrus. Alaska Cooperative Wildlife Research Unit [College, Alaska] Spec. Rept. 1, 9 + 103 p.

Brown, K. G. 1952. Observations on the newly born leopard seal. Nature 170: 982–983.

Brown, K. G. 1957. The leopard seal at Heard Island, 1951–54. Interim Repts. Australian Natl. Antarct. Res. Exped., no. 16, 2 + 34 p., 8 pls.

Brown, R. W. 1954. Composition of scientific words . . . Washington, publ. by author [geologist, U.S. Geol. Surv.], 882 + 3 p.

Bruce, W. S. 1915. Measurements and weights of antarctic seals . . . [part 11, p. 159–174, 2 pls.]. *In* Report on the scientific results of the voyage of S. Y. "Scotia" during the years 1902, 1903, and 1904 . . . Edinburgh, Scottish Oceanogr. Lab.

Brünnich, M. T. 1771. Zoologiae fundamenta praelectionibus academicis accomodata. Grunde i dyrelaeren. Hafnia et Lipsiae, 254 p. (Date from Bull. Zool. Nom. 4:307.)

Buffon, G. L. L. 1782. L'ours-marin [Supplément. . . Tome 6, p. 336–357, pl. 47]. *In* Histoire naturelle, généralle et particulière, avec la description du cabinet du Roi. Ed. 2. Paris, 37 vols. and supplement of 7 vols., 1774–89.

Burney, J. 1803–1817. A chronological history of the discoveries in the South Sea or Pacific Ocean. London, Luke Hansard, 5 vols.

Cabrera, A. 1940. Notas sobre carnívoros sudamericanos. Notas Mus. La Plata, Buenos Aires, Tomo 5, zool. 29. p. 1–22.

Cabrera, A., and J. Yepes. 1940. Mamíferos sud-americanos. Buenos Aires, Compañia Argentina de Editores, 370 p., 78 col. pls.

Carter, T. D., J. E. Hill, and G. H. H. Tate. 1945. Mammals of the Pacific world. New York, Macmillan, 16 + 227 p.

Chapin, J. P. 1936. Islands west of South America. Nat. Hist., New York, 38: 31–55.

Chapskiy, K. K. 1936. [The walrus of the Kara Sea]. Trans. Arct. Inst., Leningrad, Tom 67, 111 p. (In Russian; résumé in English, p. 112–124.)

Chapskiy, K. K. 1955*a*. [Attempt to review the systematics and diagnostics of seals of the subfamily Phocinae]. Trudy Zoologicheskogo Instituta, Tom 17, p. 160–199. (In Russian.)

Chapskiy, K. K. 1955*b*. [On the history of origin of the Caspian and Baikal seals]. Trudy Zoologicheskogo Instituta, Tom 17, p. 200–216. (In Russian.)

Chiasson, R. B. 1957. The dentition of the Alaskan fur seal. J. Mammal. 38: 310–319.

Choris, L. 1822. Port San-Francisco et ses habitants [un-numbered chapter, p. 1–10 and pls. 1–14]. *In* Voyage pittoresque autour du monde . . . Paris, folio with many p. and pls., not consec. numbered.

Clark, J. W. 1873. On the eared seals of the Auckland Islands. Proc. Zool. Soc. Lond. 1873:750–760.

Clark, J. W. 1875. On the eared seals of the islands of St. Paul and Amsterdam, with a description of the fur-seal of New Zealand, and an attempt to distinguish and re-arrange the New Zealand Otariidae. Proc. Zool. Soc. Lond. 1875:650–677, 3 pls.

Cobb, W. M. 1933. The dentition of the walrus, *Odobenus obesus*. Proc. Zool. Soc. Lond. 1933:645–668, 6 pls.

Collett, R. 1881. On *Halichoerus grypus* and its breeding on the Fro Islands off Trondhjems-fjord in Norway. Proc. Zool. Soc. Lond. 1881:380–387.

Collyer, R. D., and J. L. Baxter. 1951. Observations on pinnipeds of San Miguel Island. Calif. Fish Game, 37:511.

Conisbee, L. R. 1953. A list of the names proposed for genera and subgenera of Recent mammals from the publication of T. S. Palmer's "Index Generum Mammalium" 1904 to the end of 1951. London, Brit. Mus. (Nat. Hist.) 109 p.

Cook, J. 1785. A voyage to the Pacific Ocean undertaken, by the command of His Majesty, for making discoveries in the northern hemisphere . . . 1776–1780. London, G. Nicol and T. Cadell, v. 1 and v. 2 by Cook, v. 3 by James King.

Cope, E. D. 1896. The primary factors of organic evolution. Chicago, Open Court Pub. Co., 16 + 532 p.

Cowan, I. McT., and C. J. Guiguet. 1956. The mammals of British Columbia. Brit. Col. Provincial Mus., Handbook 11, 251 p.

Croix, P. Magne de la. 1937. De l'origine des pinnipèdes. An. Soc. Cient. Argent. 123:321–328.

Cross, C. M. P. 1928. South African sealing industry. Fur J., January, p. 24, 37–39.

Cunha Veira, C. da. 1955. Pinnipedia [p. 456]. *In* Lista remissiva dos mamíferos do Brasil. Arch. Zool. S. Paulo 8:341–474.

Cuvier, F. (see Geoffroy Saint-Hilaire, 1826).

Cuvier, G. (see Gray, 1827*b*).

Dampier, W. 1697 [1937]. A new voyage round the world . . . with an introduction by Sir Albert Gray. London, Adam and Charles Black, 42 + 376 p.

Darling, F. F. 1947. Wild life of Britain. London, Collins, 48 p.

Davis, R. H. 1955. Deep diving and submarine operations . . . Ed. 6. London, Siebe, Gorman and Co., 2 parts.

Degerbøl, M. 1950. Pinnipedia [p. 143]. *In* List of Danish vertebrates. Copenhagen, Dansk Videnskabs Forlag, 180 p., map. (In English.)

Degerbøl, M., and P. Freuchen. 1935. Mammals [vol. 2, nos. 4–5, 278 p.]. In Report of the fifth Thule expedition 1921–24. Copenhagen, Nordisk Forlag. Part I. Systematic notes, by Degerbøl. Part II. Field notes and biological observations, by Freuchen.

DeKay, J. E. 1842. Mammalia [part 1, 16 + 146 p., 33 pls.]. *In* Zoology of New-York, or the New-York fauna . . . Albany, Thurlow Weed, Printer to the State, 5 vols.

Deutsches Hydrographisches Institut. 1950. Atlas der Eisverhältnisse des Nordatlantischen Ozeans und Übersichtskarten der Eisverhältnisse des Nord- und Südpolargebietes. Hamburg, D. H. I., atlas no. 2335, 24 p., 28 pls.

Dorofeev, S. W., and S. J. Freimann (see Naumov and Smirnov, 1936).

Doutch, H. F. 1952. Rookery Island [Macquarie]. Geogr. Mag. 25:224–231.

Doutt, J. K. 1942. A review of the genus Phoca. An. Carneg. Mus. 29:61–125.

Doutt, J. K. 1954. Observations on mammals along the east coast of Hudson Bay and the interior of Ungava. An. Carneg. Mus. 33:235–249.

Downs, T. 1953. A mandible of the seal *Allodesmus kernensis* from the Kern River Miocene [Temblor] of California. Bull. S. Calif. Acad. Sci. 52:93–102.

Downs, T. 1956. A new pinniped [*Atopotarus*] from the Miocene of southern California, with remarks on the Otariidae. J. Paleont. 30:115–131, 1 pl.

Dunbar, M. J. 1949. The Pinnipedia of the arctic and subarctic. Bull. Fish. Res. Bd. Can. 85:1–22.

Dunbar, M. J. 1954. The status of the Atlantic walrus, *Odobenus rosmarus* (L.), in Canada. Arct. Circ. 8:11–14.

Dunbar, M. J. 1956. The status of the Atlantic walrus, *Odobenus rosmarus* (L.), in Canada [p. 59–61]. *In* Proc. and Pap. 5th Technical Meeting, Internatl. Union for the Protection of Nature, Copenhagen, 1954. Bruxelles, pub. by the Secretariat.

Duncan, A. 1956. Notes on the food and parasites of the grey seal, *Halichoerus grypus* (Fabricius), from the Isle of Man. Proc. Zool. Soc. Lond. 126:635–644.

Dybowski, B. 1873. Ueber die Baikal-Robbe, *Phoca baicalensis.* Arch. Anat. Physiol. Wissensch. Medizin 1873:109–125, 2 pls.

Eadie, J., and R. L. Kirk. 1952. The sodium and potassium concentration in the blood cells and plasma of the elephant seal. Aust. J. Sci. 15:26–27.

Eibl-Eibesfeldt, I. 1955. Ethologische Studien am Galápagos-Seelöwen, *Zalophus wollebaeki* Sivertsen. Z. Tierpsychol. 12:286–303.

Ekman, S. 1953. Zoogeography of the sea. London, Sidgwick and Jackson, 14 + 417 p.

Ellerman, J. R., and T. C. S. Morrison-Scott. 1951. Checklist of Palaearctic and Indian mammals 1758 to 1946. London, Brit. Mus. (Nat. Hist.) 810 p.

Ellerman, J. R., T. C. S. Morrison-Scott, and R. W. Hayman. 1953. Southern African mammals 1758–1951: a reclassification. London, Brit. Mus. (Nat. Hist.), 363 p., 3 maps.

Elliot, D. G. 1901. Order Pinnipedia [p. 355–365, 12 pls.]. *In* A synopsis of the mammals of North America and the adjacent seas. Publ. Field Mus., zool. ser., vol. 2.

Elliot, D. G. 1904. Pinnipedia [part 2, p. 538–546, pls. 53–66]. *In* The land and sea mammals of Middle America and the West Indies. Publ. Field Mus. 95, zool, ser., vol. 4, 2 parts.

Elliott, H. F. I. 1953. The fauna of Tristan da Cunha. Oryx 2:41–53.

Engle, E. T. 1926. The intestinal length in Steller's sea lion. J. Mammal. 7:28–30.

Erdbrink, D. P. 1953. A review of fossil and recent bears of the Old World . . . Deventer, DeLange, 2 parts, 12 + 597 p. and index.

Erxleben, J. C. P. 1777. Systema regni animalis per classes, ordines, genera, species . . . Classis I. Mammalia. Lipsiae, 48 + 636 + 64 p.

Eschricht, D. F. 1866. On the species of the genus *Orca* inhabiting the northern seas [p. 151–188]. *In* Recent memoirs on the Cetacea, by Eschricht *et al.* London, Ray Society, publ. 40, 312 p., 6 pls.

Fabricius, O. 1791. Udfarlig beskrivelse over de Gronlandske saele. Skrivter af Naturhistorie-Selskabet, Kjøbenhavn, Bind 1, Hefte 2, p. 73–170, 2 pls.

Falla, R. A. 1953. Southern seals: population studies and conservation problems. Proc. 7th Pac. Sci. Congr. 4:706.

Fauvel, A. 1892. La faune du Chan-Toung. Rev. Quest. Sci., Tome 1, 463 p. (not seen; quoted by Leroy, 1940).

Fay, F. H. 1957. History and present status of the Pacific walrus population. Trans. 22d N. Amer. Wildl. Conf., 431–445.

Fetcher, E. S. 1939. The water balance in marine mammals. Quart. Rev. Biol. 14:451–459.

Finnegan, Susan. 1934. On a new species of mite of the family Halarachnidae from the southern sea lion. Discovery Repts. 8:319–328.

Fisher, H. D. 1954. Rapid preparation of tooth sections for age determinations. J. Wildl. Manage. 18:535–537.

Fisher, H. D. 1956. Utilization of Atlantic harp seal populations. Trans. 20th N. Amer. Wildl. Conf., 507–518.

Fisher, J. 1954. Evolution and bird sociality [p. 71–83]. *In* Evolution as a process. Ed. by Julian Huxley. London, George Allen and Unwin, 8 + 367 p.

Fleming, J. 1822. Palmata [= Pinnipedia plus *Lutra* and *Enhydra,* vol. 2,

p. 186–188]. *In* The philosophy of zoology . . . Edinburgh, Archibold Constable, 2 vols.

Fleming, J. 1828. A history of British animals . . . Edinburgh, 2 vols.

Flower, W. H. 1884. Pinnipedia [part 2, p. 186–218]. *In* Catalogue of the specimens illustrating the osteology and dentition of vertebrated animals . . . in the museum of the Royal College of Surgeons of England. London, J. and A. Churchill, 2 parts.

Follett, W. I. 1955. An unofficial interpretation of the international rules of zoological nomenclature. "Issued with the cooperation of the California Academy of Sciences and the Society of Systematic Zoology . . . September 1955 / not published," 5 + 99 p.

Forster, G. 1777. A voyage round the world, in H.B.M. Sloop, "Resolution," commanded by Capt. James Cook, during the years 1772, 3, 4, and 5. London, 2 vols.

Forster, J. R. 1844. Descriptiones animalium quae in itinerere ad maris australis terras per annos [1772–74] suscepto . . . ed. Henrico Lichtenstein. Berolini, Koeniglich-Preussische Akad. d. Wissenschaften, 13 + 424 p.

Fowler, S. 1947. A landing on Pedra Branca. Proc. Roy. Zool. Soc. New S. Wales 1946–47:22–26.

Fraser, F. C. 1935. Zoological notes from the voyage of Peter Mundy, 1655–56 . . . sea elephant on St. Helena . . . Proc. Linn. Soc. Lond. 1934–35:32–37, 1 pl.

Fraser, F. C. 1940. Three anomalous dolphins from Blacksod Bay, Ireland. Proc. Roy. Irish Acad. 45 B:413–455, 7 pls.

Frechkop, S. 1955. Ordre des pinnipèdes [Tome 17, Fasc. 1., p. 292–340]. *In* Traité de zoologie . . . publie sous la direction de Pierre-P. Grassé. Paris, Masson, 18 vols.

Freuchen, P. (see Degerbøl and Freuchen, 1935).

Freund, L. 1933. Pinnipedia [Lief. 24, Teil 12, p. k_2–k_{83}]. *In* Die Tierwelt der Nord-und Ostsee, bergründet von G. Grimpe und E. Wagler . . . Leipzig, Akademische Verlagsgesellschaft, many parts.

Friant, Madeleine. 1956. Morphologie et developpement du cerveau des mammifères eutheriens. I. Séries des insectivores et des carnassiers. An. Soc. Zool. Belg., Tome 86, 1955–56, Fasc. 2:249–279.

Frizzell, D. L. 1933. Terminology of types. Amer. Midl. Nat. 14:637–668.

Fry, D. H., Jr. 1939. A winter influx of sea lions from Lower California. Calif. Fish Game 25:245–250.

Gama, Maria Manuela da. 1957. Pinnipedia [p. 202–206]. *In* Mamíferos de Portugal (chaves para a sua determinação). Mem. Mus. Zool. Univ. Coimbra 246:1–246.

Geoffroy Saint-Hilaire, É., and F. Cuvier. 1826. Phoques [vol. 39, p. 540–559]. *In* Dictionnaire des sciences naturelles . . . Strasbourg and Paris, 60 vols. + atlas of 12 vols., 1816–30.

Gervais, H., and F. Ameghino. 1880. Les mammifères fossiles de l'Amérique du Sud. Buenos Aires and Paris, 12 + 225 p. (alternate p. in Spanish and French).

Gibbney, L. F. 1957. The seasonal reproductive cycle of the female elephant seal . . . at Heard Island. Repts. Australian Natl. Antarc. Res. Exped., ser. B, vol. 1, p. 4 + 26.

Giebel, C. G. 1847 [1848]. Phoca [Sect. 3, p. 284–292]. *In* Allgemeine Encyclopädie der Wissenschaften und Künste . . . von J. S. Ersch und J. G. Gruber. Leipzig.

Gijzen, Agatha. 1956. Les pinnipèdes. Zoo (Soc. Roy. Zool. Anvers), Année 22, no. 1, p. 3–37.

Gill, T. 1866*a*. Prodrome of a monograph of the pinnipedes. Proc. Essex Inst., Salem, Communications, vol. 5:3–13.

Gill, T. 1866*b*. On a new species of the genus *Macrorhinus*. Proc. Chicago Acad. Sci. 1:33–34.

Gill, T. 1872. Arrangement of the families of mammals with analytical tables. Smithson. Misc. Coll. no. 230, vol. 11:6 + 98 p.

Gill, T. 1873. The ribbon seal of Alaska. Amer. Nat. 7:178–179.

Gill, T. 1897. The distribution of marine mammals. Science, n.s., vol. 5, no. 129:955–956.

Gistel [or Gistl], J. von N. F. X. 1848. Naturgeschichte des Thierreichs für höhere Schulen. Stuttgart, 16 + 216 + 4 p., 32 pls.

Gmelin, J. F. 1788. Mammalia [Tome 1, Part 1, 10 + 232 p.]. *In* Systema naturae . . . editio decima tertia, aucta, reformata, cura J. F. Gmelin. Lipsiae, 3 vols. in 7, 1788–93.

González Ruiz, G. T. 1955. La fauna del litoral Atlantico: lobos y elefantes marinos, pingüinos y aves guaneras. Natura, Buenos Aires, 1:121–129.

Goodwin, G. G. 1954. Southern records for arctic mammals and a northern record for Alfaro's rice rat. J. Mammal. 35:258.

Gray, J. E. 1825. An outline of an attempt at the disposition of Mammalia into tribes and families, with a list of the genera apparently appertaining to each tribe. Ann. Phil., n.s., vol. 10 (vol. 26 of the whole series):337–344.

Gray, J. E. 1827. Mammalia [vol. 5, 391 p.]. *In* The animal kingdom arranged in conformity with its organization, by the Baron [G.] Cuvier, with additional descriptions . . . by Edward Griffith . . . and others. London, George B. Whittaker, 16 vols., 1827–35.

Gray, J. E. 1828. Spicilegia zoologica; or original figures and short systematic descriptions of new and unfigured animals. London, 2 parts (1828 and 1830), each with pls.

Gray, J. E. 1837. Description of some new or little known Mammalia, principally in the British Museum collection. Mag. Nat. Hist., n.s., 1:577–587.

Gray, J. E. 1843. List of the specimens of Mammalia in the collection of the British Museum. London, Brit. Mus. 28 + 216 p.

Gray, J. E. 1844, 1875. The seals of the southern hemisphere [p. 1–8, pls. 1–10, 14–17, *1844*; p. 9–12, *1875*]. *In* The zoology of the voyage of H.M.S. "Erebus" & "Terror" . . . during the years 1839 to 1843 . . . edited by John Richardson . . . and John Edward Gray . . . London, 2 vols.

Gray, J. E. 1847. List of the osteological specimens in the collection of the British Museum. London, Brit. Mus., 25 + 147 p.

Gray, J. E. 1849. On the variation in the teeth of the crested seal, *Cystophora cristata*, and on a new species of the genus from the West Indies. Proc. Zool. Soc. Lond. 1849:91–93.

Gray, J. E. 1850. Seals [part 2, p. 5 + 48]. *In* Catalogue of the specimens of Mammalia in the collection of the British Museum. London, Brit. Mus., 3 parts.

Gray, J. E. 1859. On the sea-lions, or lobos marinos of the Spaniards, on the coast of California. Proc. Zool. Soc. Lond. 1859:357–361, 1 pl.

Gray, J. E. 1864. Notes on the seals (Phocidae), including the description of a new seal (*Halicyon richardii*), from the west coast of North America. Proc. Zool. Soc. Lond. 1864:27–34.

Gray, J. E. 1866*a*. Notes on the skulls of sea-bears and sea-lions (Otariadae) in the British Museum. Ann. Mag. Nat. Hist., ser. 3, 18:228–237.

Gray, J. E. 1866*b*. Pinnipedia [p. 1–60, figs. 1–18]. *In* Catalogue of seals and whales in the British Museum. London, Brit. Mus. 8 + 402 p.

Gray, J. E. 1869. Notes on seals (Phocidae) and the changes in the form of their lower jaw during growth. Ann. Mag. Nat. Hist., ser. 4, 4:342–346.

Gray, J. E. 1871. Supplement to the catalogue of seals and whales in the British Museum. London, Brit. Mus. 103 p.

Gray, J. E. 1872. Description of the younger skull of Steller's sea-bear (*Eumetopias stelleri*). Proc. Zool. Soc. Lond. 1872:737–743.

Gray, J. E. 1874. Hand-list of seals, morses, sea-lions, and sea-bears in the British Museum. London, Brit. Mus. 44 p., 30 pls.

Gray, J. E. 1875. (see Gray, 1844).

Gregory, W. K., and M. Hellman. 1939. On the evolution and major classification of the civets (Viverridae) and allied fossil and recent Carnivora: a phylogenetic study of the skull and dentition. Proc. Amer. Phil. Soc. 81: 309–392, frontis.

Grevé, C. 1896. Die geographische Verbreitung der Pinnipedia. Nova Acta Abh. der Kaiserl. Leop.-Carol. Deutschen Akad. der Naturforscher, Band 66, no. 4:287–332, 4 pls. (maps in color).

Grimm, O. 1883. Pinnipedia [p. 44–47]. *In* Fishing and hunting on Russian waters. St. Petersburg, prepared for Internatl. Fish. Exhibn. London, 1883, 55 p. (In English.)

Gwynn, A. M. 1953*a*. The status of the leopard seal at Heard Island and Macquarie Island, 1948–1950. Interim Repts. Australian Natl. Antarct. Res. Exped., no. 3, 33 p.

Gwynn, A. M. 1953*b*. Notes on the fur seals [and sea-lion] at Macquarie Island and Heard Island. Interim Repts. Australian Natl. Antarct. Res. Exped., no. 4, 16 p.

Hall, E. R. 1940. Pribilof fur seal on California coast. Calif. Fish Game 26: 76–77.

Hall, E. R. 1946. Mammals of Nevada. Berkeley and Los Angeles, Univ. Calif. Press, 11 + 710 p.

Hall, T. S. 1903. [Crabeater seals in Australian waters]. Nature 67:327–328.

Hamilton, J. E. 1934. The southern sea lion, *Otaria byronia* (de Blainville). Discovery Repts. 8:269–318, 13 pls.

Hamilton, J. E. 1939*a*. The leopard seal *Hydrurga leptonyx* (de Blainville). Discovery Repts. 18:239–264, 7 pls.

Hamilton, J. E. 1939*b*. A second report on the southern sea lion, *Otaria byronia* (de Blainville). Discovery Repts. 19:121–164, 8 pls.

Hamilton, J. E. 1940. On the history of the elephant seal, *Mirounga leonina* (Linn.). Proc. Linn. Soc. Lond. 1939–40:33–37.

Hamilton, J. E. 1945. The Weddell seal in the Falkland Islands. Proc. Zool. Soc. Lond. 114:549.

Hanna, G. D. 1923. Rare mammals of the Pribilof Islands, Alaska. J. Mammal. 4:209–215, 1 pl.

Hanna, G. D. 1926. Expedition to the Revillagigedo Islands, Mexico, in 1925. Proc. Calif. Acad. Sci., ser. 4, vol. 15, 113 p., 10 pls.

Harper, F. 1956. The mammals of Keewatin. Misc. Publ. Univ. Kansas Mus. Nat. Hist. 12:1–94, 6 pls., 1 map.

Harrison, R. J., L. H. Matthews, and J. M. Roberts. 1952. Reproduction in some Pinnipedia. Trans. Zool. Soc. Lond. 27:437–540, 4 pls.

Harrison, R. J., and J. D. W. Tomlinson. 1956. Observations on the venous system in certain Pinnipedia and Cetacea. Proc. Zool. Soc. Lond. 126: 205–233.

Havinga, B. 1933. Der Seehund (*Phoca vitulina* L.) in den Holländischen Gewässern. Tijdschr. Ned. Dierk. Ver., Ser. 3, Aft. 2–3, p. 79–111.

Hay, O. P. 1930. Pinnipedia [vol. 2, p. 555–565]. *In* Second bibliography and catalogue of the fossil Vertebrata of North America. Publ. Carneg. Instn. 390, 2 vols., 1929–30.

Hayes, J. G. 1928. Seals (Pinnipedia) [p. 103–108]. *In* Antarctica, a treatise on the southern continent. London, Richards, 15 + 448 p., 16 pls.

Hediger, H. 1955. Studies of the psychology and behaviour of captive animals in zoos and circuses. London, Butterworth, 8 + 166 p., 16 pls.

Heller, E. 1904. Mammals of the Galapagos Archipelago, exclusive of the Cetacea. Pap. Hopkins Stanford Galapagos Exped., 1898–1899. Proc. Calif. Acad. Sci., ser. 3 (zool.), vol. 3:233–250, 1 pl.

Hemming, F. 1955. Second report on the status of the generic names "*Odobenus*" Brisson, 1762, and "*Rosmarus*" Brünnich, 1771 (class Mammalia) . . . Bull. Zool. Nom. 11:196–198.

Henry, T. R. [1951]. The white continent. London, Eyre and Spottiswoode, 13 + 211 p.

Hentschel, E. 1937. Naturgeschichte der nordatlantischen Wale und Robben [Heft 1, 6 + 54 p., 10 pls.]. *In* Handbuch der Seefischerei nordeuropas, Band III, Systematik u. Biologie . . . Stuttgart, Erwin Nägele.

Hermann, J. 1779. Beschreibung der Münchs-Robbe. Beschäf. Berlin Ges. Naturf. Freunde 4:456–509, 2 pl.

Hewer, H. R. 1957. A Hebridean breeding colony of grey seals *Halichoerus grypus* (Fab.), with comparative notes on the grey seals of Ramsey Island, Pembrokeshire. Proc. Zool. Soc. Lond. 128:23–66, 2 pls.

Hickling, Grace (Mrs.) 1956. The grey seals of the Farne Islands. Trans. Nat. Hist. Soc. Northumb., Durham, and Newcastle upon Tyne, n.s., 11:

230–244. (In 1957 another article, with the same title and by the same author, appeared in this journal, 12:93–133, 5 pls.)

Holdgate, M. W., R. W. LeMaitre, M. K. Swailes, and N. M. Wace. 1956. The Gough Island scientific survey, 1955–56. Nature 178:234–236.

Hombron (see C. H. Jacquinot, 1841).

Hopkins, G. H. E. 1949. The host-associations of the lice of mammals. Proc. Zool. Soc. Lond. 119:387–604.

Howard, Patricia. 1954. A.N.A.R.E. bird banding and seal marking. Vict. Nat., Melb. 71:73–82.

Howell, A. B. 1929. Contributions to the comparative anatomy of the eared and earless seals (genera *Zalophus* and *Phoca*). Proc. U.S. Nat. Mus. no. 2736, vol. 73, art. 15:1–142, 1 pl.

Howell, A. B. 1930. Aquatic mammals, their adaptations to life in the water. Baltimore, Charles C Thomas, 338 p.

Hubbs, C. L. 1956*a*. Back from oblivion. Pacific Discovery 9:14–21.

Hubbs, C. L. 1956*b*. The Guadalupe fur seal still lives! Zoonooz, San Diego, 29:6–9.

Huber, E. 1934. Anatomical notes on Pinnipedia and Cetacea. Publ. Carneg. Instn. 447, part 4, p. 105–136.

Huey, L. M. 1942. Pribilof fur seal taken in San Diego County, California. J. Mammal. 23:95–96.

Huxley, J. 1954. The evolutionary process [p. 1–23]. *In* Evolution as a process, ed. by Huxley. London, George Allen and Unwin, 8 + 367 p.

Illiger, C. 1811. Prodromus systematis mammalium et avium; additis terminis zoographicis utriusque classis, eorumque versione germanica. Berlin, C. Salfeld, 18 + 301 + 1 p.

Illiger, J. C. W. 1815. Ueberblick der Säugethiere nach ihrer Vertheilung über die Welttheile, Abhandl. Physik. Klasse der Königlich-Preuss. Akad. Wissensch. 1804–1811, p. 39–159.

Imler, R., and H. R. Sarber. 1947. Harbor seals and sea lions in Alaska. U.S. Dept. Interior, Fish and Wildlife Service, Spec. Sci. Rept. 28, 22 p.

Inukai, T. 1942 [Hair seals (azarashi) in our northern waters]. Shokubutsu Dobutsu (Botany and Zoology), vol. 10, no. 10:927–932; vol. 10, no. 11: 1025–1030. (In Japanese; see Scheffer, 1956.)

Iredale, T., and E. LeG. Troughton. 1934. Order Pinnipedia [p. 87–89]. *In* A check-list of the mammals recorded from Australia. Mem. Aust. Mus., Sydney, no. 6, 12 + 122 p.

Irving, L. 1942. The action of the heart and circulation during diving. Trans. N.Y. Acad. Sci. ser. 2, vol. 5:11–16.

Irving, L., K. C. Fisher, and F. C. McIntosh. 1935. The water balance of a marine mammal, the seal. J. Cell. Comp. Physiol. 6:387–391.

Isachsen, F. 1933. Verdien av den norske Klappmysfangst langs Sydøstgrønland. Norges Svalbard-og Ishavs-Undersøkelser, Oslo, Meddelelse 22:1–24. (English summary.)

Jacquinot, C. H. [1841] 1842–54. Voyage au Pôle Sud et dans l'Océanie sur les corvettes "l'Astrolabe" et "la Zélée," executé . . . pendant . . . 1837–1840 sous le commandement de J. Dumont d'Urville . . . publié . . . sous la direction supérieure de M. Jacquinot . . . Paris, 30 vols. in 27 and atlas. (Zoological articles are by J. -B. Hombron and Honore [not C. H.] Jacquinot, *circa* 1842; and by Honore Jacquinot and Jacques Pucheran, 1853. For full description see Cat. Libr. Brit. Mus. (Nat. Hist.) 1904, vol. 2, E-K, p. 605).

Jaeger, E. C. 1955. A source-book of biological names and terms. Springfield, Illinois, Charles C Thomas, 35 + 317 p.

Jentink, F. A. 1892. Catalogue systématique des mammifères . . . [vol. 11, 219 p.]. *In* Revue méthodique et critique des collections. Leiden, Museum d'Histoire Naturelle des Pays-Bas [Rijksmuseum van Natuurlijke Historie], 14 vols. in 13, 1862–1908.

Johnson, G. L. 1893. Observations on the refraction and vision of the seal's eye. Proc. Zool. Soc. Lond. 1893:719–723.

Jordan, D. S., and G. A. Clark. 1898. The history, condition, and needs of the herd of fur seals resorting to the Pribilof Islands [part 1, 7 + 249 p., 12 pls.]. *In* The fur seals and fur-seal islands of the North Pacific Ocean . . . by Jordan *et al.* Washington, Govt. Print. Off., Treasury Dept. Doc. 2017, 4 parts.

Jordan, D. S., and G. A. Clark. 1899. The species of *Callorhinus* or northern fur seal [part 3, p. 2–4]. *In* The fur seals . . . (see above).

Kelemen, G., and A. Hasskó. 1931. Das Stimmorgan des Seelöwen (*Otaria jubata* [= *Zalophus*]. Z. Anat. EntwGesch. 95:497–511.

Kellogg, R. 1922. Pinnipeds from Miocene and Pleistocene deposits of California . . . and a résumé of current theories regarding origin of Pinnipedia. Bull. Dep. Geol. Univ. Calif. 13:23–132.

Kellogg, R. 1925. New pinnipeds from the Miocene diatomaceous earth near Lompoc, California [part 4, p. 71–95]. *In* Additions to the Tertiary history of the pelagic mammals on the Pacific coast of North America. Contrib. Palaeont. Carneg. Instn. 348, 5 parts.

Kellogg, R. 1931. Pelagic mammals from the Temblor formation of the Kern River region, California. Proc. Calif. Acad. Sci., ser. 4, 19:217–397.

Kellogg, R. 1936. Resemblances of the Archaeoceti to the Pinnipedia [p. 311–319]. *In* A review of the Archaeoceti. Publ. Carneg. Instn. 482, 15 + 366 p., 37 pls.

Kellogg, R. 1942. Tertiary, Quaternary, and Recent marine mammals of South America and the West Indies. Proc. 8th Amer. Sci. Congr., Washington, 1940, 3:445–473.

Kenyon, K. W. 1952. Diving depths of the Steller sea lion and Alaska fur seal. J. Mammal. 33:245–246.

Kenyon, K. W., and V. B. Scheffer. 1955. The seals, sea-lions, and sea otter of the Pacific coast. U.S. Dept. Interior, Fish and Wildlife Service, Circ. 32, 2 + 34 p.

Kenyon, K. W., V. B. Scheffer, and D. G. Chapman. 1954. A population study of the Alaska fur-seal herd. U.S. Dept. Interior, Fish and Wildlife Service, Spec. Sci. Rept. Wildl. 12, 77 p.

Kenyon, K. W., and F. Wilke. 1953. Migration of the northern fur seal, *Callorhinus ursinus.* J. Mammal. 34:86–98.

Kettlewell, H. B. D., and R. Rand. 1955. Elephant seal cow and pup on South African coast. Nature 175:1000–1001.

King, Judith E. 1954. The otariid seals of the Pacific coast of America. Bull. Brit. Mus. (Nat. Hist.), Zool. 2:309–337, 2 pls.

King, Judith E. 1956. The monk seals (genus *Monachus*). Bull. Brit. Mus. (Nat. Hist.), Zool. 3:201–256, 6 pls.

Kumlien, L. 1879. Contributions to the natural history of arctic America, made in connection with the Howgate polar expedition, 1877–78. Bull. U.S. Nat. Mus. 15:1–179.

Kuroda, N. 1938. A list of the Japanese mammals. Tokyo, pub. by author, 4 + 122 p.

Kuroda, N. 1940. A monograph of the Japanese mammals. Tokyo and Osaka, Sanseido Co., 16 + 311 p., 48 col. pls. (In Japanese, with Latin and English names of mammals.)

Law, P. G., and T. Burstall. 1953. Heard Island. Interim Repts. Australian Natl. Antarc. Res. Exped. no. 7, 32 p.

Laws, R. M. 1952. A new method of age determination for mammals. Nature 169:972–974.

Laws, R. M. 1953*a*. The seals of the Falkland Islands and Dependencies. Oryx 2:87–97.

Laws, R. M. 1953*b*. The elephant seal (*Mirounga leonina,* Linn.). I. Growth and age. Falkland Is. Depend. Surv. Sci. Repts. 8:1–62, 5 pls.

Laws, R. M. 1953*c*. A new method of age determination in mammals with special reference to the elephant seal *Mirounga leonina,* Linn. Falkland Is. Depend. Surv. Sci. Repts. 2:1–12, 1 pl.

Laws, R. M. 1956*a*. The elephant seal (*Mirounga leonina,* Linn.). II. General, social and reproductive behaviour. Falkland Is. Depend. Surv. Sci. Repts. 13:1–88, 7 pls.

Laws, R. M. 1956*b*. The elephant seal (*Mirounga leonina,* Linn.). III. The physiology of reproduction. Falkland Is. Depend. Surv. Sci. Repts. 15:1–66, 5 pls.

Laws, R. M. 1956*c*. Growth and sexual maturity in aquatic mammals. Nature 178:193–194.

Laws, R. M. 1957. On the growth rates of the leopard seal, *Hydrurga leptonyx* (De Blainville, 1820). Säugetierk. Mitt. 5:49–55.

Laws, R. M., and R. J. F. Taylor. 1957. A mass dying of crabeater seals, *Lobodon carcinophagus* (Gray). Proc. Zool. Soc. Lond. 129:315–324.

Leidy, J. 1853. [*Stenorhynchus vetus* n. sp.]. Proc. Acad. Nat. Sci. Philad. 6:377.

Leone, C. A., and A. L. Wiens. 1956. Comparative serology of carnivores. J. Mammal. 37:11–23.

Lepechin, I. 1778. Phocarum species descriptae. Acta Academiae Scientiarum Imperialis Petropolitanae, 1777, vol. 1, p. 257–266, 4 pls.

Leroy, P. 1940. On the occurence [*sic*] of a hair-seal, *Phoca Richardsi* (Gray), in the coast of North China. Bull. Fan Memorial Institute of Biology, zool. ser., 10:61–68, 1 pl.

LeSouef, A. S. 1929. Occurrence of the crab-eating seal *Lobodon carcinophaga* Hombron and Jacuinot [*sic*], in New South Wales. Aus. Zool. 6:99, 1 pl.

Lesson, R.-P. 1826. Sur le phoque léopard de mer (sea leopard) des Orcades australes; par James Weddell. (Voy. towards the south pole, etc. . . .) Reviewed by Lesson. [Ferussac's] Bulletin des Sciences Naturelles et de Géologie, Tome 7:437–438.

Lesson, R.-P. 1827. Manuel de mammalogie, ou histoire naturelle des mammifères. Paris, Roret, 15 + 441 + 3 p.

Lesson, R.-P. 1828. Phoque [vol. 13, p. 400–426, January 1828]. *In* Dictionnaire classique d'histoire naturelle, ed. by Baron J. B. G. M. Bory de Saint-Vincent. Paris, Rey et Gravier, 17 vols., 1822–31.

Lewis, F. 1942. Notes on Australian seals. Vict. Nat., Melb., 59:24–26.

Lillie, H. R. 1956. The hood seal (*Cystophora cristata*) [p. 62–63]. *In* Pap. and Proc. 5th Technical Meeting Internatl. Union for the Protection of Nature, Copenhagen, 1954. Bruxelles, pub. by the Secretariat.

Lindsey, A. A. 1937. The Weddell seal in the Bay of Whales, Antarctica. J. Mammal. 18:127–144.

Lindsey, A. A. 1938. Notes on the crab-eater seal. J. Mammal. 19:459–461.

Linnaeus, C. 1758. Systema naturae . . . , Ed. 10, revised. Stockholm, Laurentii Salvii, vol. 1, part 1, 532 p.

Lockley, R. M. 1954. The Atlantic grey seal. Oryx 2:384–387.

Loomis, F. B. 1911. A new mink from the shell heaps of Maine. Amer. J. Sci., ser. 4, 31:227–229.

Luke, H. 1953. Robinson Crusoe's two islands [Juan Fernández and Tobago]. Geogr. Mag. 26:291–297.

Lydekker, R. 1909. On the skull-characters in the southern sea-elephant. Proc. Zool. Soc. Lond. 1909:600–606.

M'Bain, J. 1858. On the skull of a seal (*Otaria Gillespii,* M'Bain) from the Gulf of California; with some preliminary observations on the amphibious Carnivora. Proc. R. Phys. Soc. Edinb. 1854–58, 1:422–428.

Mackintosh, N. A., and H. F. P. Herdman. 1940. Distribution of the pack-ice in the Southern Ocean. Discovery Repts. 19:285–296, 27 pls.

Mann Fischer, G. 1955. Las aves guaneras y las posibilidades de incrementar la producción de guano blanco en Chile. Rev. Chil. Hist. Nat. no. 16, año 54, p. 1–37.

Margolis, L. 1954. List of the parasites recorded from sea mammals caught off the west coast of North America. J. Fish. Res. Bd. Can. 11:267–283.

Margolis, L. 1956. Parasitic helminths and arthropods from Pinnipedia of the Canadian Pacific coast. J. Fish. Res. Bd. Can. 13:489–505.

Markowski, S. 1952. The cestodes of seals from the antarctic. Bull. Brit. Mus. (Nat. Hist.), Zool. 1:125–150.

Marr, J. W. S. 1956. *Euphausia superba* and the antarctic surface currents; an advance note on the distribution of whale food. Norsk Hvalfangsttid. 1956: 127–134.

Matheson, C. 1950. Longevity in the grey seal. Nature 166:73–74.

Matschie, P. 1905. Eine Robbe von Laysan. S. B. Ges. Naturfr. Fr. Berl. 1905: 254–262.

Matthews, L. H. 1929. The natural history of the elephant seal. Discovery Repts. 1:234–255, 6 pls.

Matthews, L. H. 1952. Sea elephant. London, Macgibbon and Kee, 190 p., 14 pls., 1 map.

Mayer, W. V. 1949*a*. Catalogue of type specimens of mammals in the Natural History Museum of Stanford University. Proc. Calif. Zool. Club 1:29–32.

Mayer, W. V. 1949*b*. The type specimens of the Pribilof fur seal, *Callorhinus alascanus* Jordan and Clark. J. Mammal. 30:316–317.

Mayr, E. 1954. Change of genetic environment and evolution [p. 157–180]. *In* Evolution as a process, ed. by Julian Huxley. London, George Allen and Unwin, 8 + 367 p.

Mayr, E., E. G. Linsley, and R. L. Usinger. 1953. Methods and principles of systematic zoology. New York, McGraw-Hill, 9 + 328 p.

McEwen, E. H. 1954. A sporadic occurrence of an Alaskan fur seal. J. Mammal. 35:444.

McLaren, I. A. 1956. [Summary of the biology of the ringed seal in waters of southwest Baffin Island, p. 185–186]. *In* The "Calanus" expeditions in the Canadian arctic, by M. J. Dunbar. Arctic 9:178–190.

Meinertzhagen, R. 1955. The speed and altitude of bird flight (with notes on other animals). Ibis 97:81–117.

Merriam, C. H. 1897. A new fur seal or sea-bear (*Arctocephalus townsendi*) from Guadalupe Island, off Lower California. Proc. Biol. Soc. Wash. 11:175–178.

Mielche, H. 1953. Round the world with "Galathea." London, William Hodge, 241 p.

Miller, G. S., Jr. 1917. A hooded seal in Florida. Proc. Biol. Soc. Wash. 30:121.

Miller, G. S., Jr. 1924. Pinnipedia [p. 162–167]. *In* List of North American Recent mammals, 1923. Bull. U.S. Nat. Mus. 128, 16 + 673 p.

Miller, G. S., Jr. 1932. Some names applied to seals by Dybowski in 1929. Proc. Biol. Soc. Wash. 45:149–50.

Miller, G. S., Jr., and R. Kellogg. 1955. Pinnipedia [p. 782–790]. *In* List of North American Recent mammals. Bull. U.S. Nat. Mus. 205, 12 + 954 p.

Miller, M. E. 1952. Guide to the dissection of the dog. Ed. 3, reprinted 1952. Ithaca, publ. by author, 427 p.

Miller, W. C. S. 1888. The myology of the Pinnipedia [vol. 26, no. 2, p. 139–240; appendix to Turner's report]. *In* Report on the scientific results of the voyage of H.M.S. "Challenger" . . . Edinburgh, Challenger Office, 1880–95.

Mivart, St. G. 1885. Notes on the Pinnipedia. Proc. Zool. Soc. Lond. 1885:484–500.

Mohr, Erna. 1941. Ein neuer westpazifischer Seehund. Zool. Anz. 133:49–60.

Mohr, Erna. 1952*a*. Beiträge zur Kenntnis der Mähnenrobben. Zool. Gart. 19 (Portig-Festheft): 98–112.

Mohr, Erna. 1952*b*. Die Robben der europäischen Gewässer. Frankfurt am Main, Paul Schöps, Monographien der Wildsäugetiere, Band 12, 283 p., 40 pls.

Mohr, Erna. 1956. Das Verhalten der Pinnipedier [Band 8, Teil 10, p. 1–20]. *In* Handbuch der Zoologie. Berlin, Walter de Gruyter.

Montagna, W., and R. J. Harrison. 1957. Specializations in the skin of the seal (*Phoca vitulina*). Am. J. Anat. 100:81–114.

Morrell, B. 1832. A narrative of four voyages . . . from the year 1822 to 1831. New York, J. and J. Harper, 492 p.

Morrison-Scott, T. C. S. 1952. A list of British mammals. London, Brit. Mus. (Nat. Hist.), 24 p.

Müller, E. 1940. Zur Anatomie des Robbenherzens . . . Gegenbaurs Jb. 85: 59–90.

Müller, O. F. 1776. Zoologiae Danicae prodromus, seu animalium Daniae et Norvegiae indigenarum . . . Havniae, p. 32 + 282. (Has list, supplied by Otho Fabricius, of names of Greenland animals, p. viii. This list is discussed by Allen, 1880, p. 425; most modern writers regard the names as *nomina nuda.*)

Murie, O. J. 1936. Notes on the mammals of St. Lawrence Island [app. 3, p. 335–346]. *In* Archaeological investigations at Kukulik . . . by O. W. Geist and F. G. Rainey. Misc. Publs. Univ. Alaska, vol. 2.

Murphy, R. C. 1913. The trachea of *Ogmorhinus* [= *Hydrurga*], with notes on other soft parts. Bull. Amer. Mus. Nat. Hist. 32:505–506.

Murphy, R. C. 1914. Notes on the sea elephant, *Mirounga leonina* (Linné). Bull. Amer. Mus. Nat. Hist. 33:63–79, 7 pls.

Murphy, R. C. 1920. The seacoast and islands of Peru. Brooklyn Mus. Quart. 7:69–95.

Nansen, F. 1925. Hunting and adventure in the arctic. New York, Duffield, 3 + 462 p., pls.

National Research Council. 1956. Handbook of biological data, ed. by William S. Spector. Washington, Nat. Res. Council, 36 + 584 p.

Naumov, S. P. 1933. [The seals of the U.S.S.R.; the raw material basis of the marine mammal fishery, p. 1–105]. *In* Economically exploited animals of the U.S.S.R., ed. by N. A. Bobrinskoi. Moscow and Leningrad, All-Union Cooperative United Publishing House. (In Russian; English transl. in Oxford Bur. Animal Population.)

Naumov, S. P. 1941. [Pinnipedia of the Okhotsk Sea (chiefly southern part)]. Uchenye Zapiski Moskovskogo Gosudarstvennogo Pedagogicheskogo Instituta, Tom 24, no. 2:19–74. (In Russian, English summary.)

Naumov, S. P., and N. A. Smirnov. 1936. [Notes on the systematic position and geographic distribution of the Phocidae of the North Pacific Ocean, vol. 3, p. 161–184]. *In* The marine mammals of U.S.S.R. Far East (resources and commercial use), ed. by S. W. Dorofeev and S. J. Freimann. Trans. Inst. Mar. Fish. U.S.S.R. (VNIRO). (In Russian.)

Neave, S. A. 1939–40, 1950. Nomenclator zoologicus. A list of the names of genera and subgenera in zoology from the tenth edition of Linnaeus 1758 to the end of 1935. Vols. 1–4 (1939–40). Vol. 5 (1950) covers years 1936–45 with add. and corrig. to vols. 1–4. Zool. Soc. Lond.

Negus, V. E. 1949. The comparative anatomy and physiology of the larynx. London, Heinemann, 20 + 230 p.

Nehring, A. 1887*a*. Ueber eine Pelzrobben-Art von der Küste Süd-Brasiliens. Arch. Naturgesch. 1:75–94, 1 pl.

Nehring, A. 1887*b*. Über die südbrasilienische Pelzrobbe. S. B. Ges. Naturf. Fr. Berlin 1887:142–143.

Nehring, A. 1887*c*. Über eine Pelzrobbe von Rio de Janeiro [largely quoting from letter from Dr. Göldi]. S. B. Ges. Naturf. Fr. Berlin 1887:207–208.

Nelson, E. W. 1899. Mammals of the Tres Marias Islands [p. 15–19]. *In* Natural history of the Tres Marias Islands, Mexico. N. Amer. Fauna 14.

Newell, I. M. 1947. Studies on the morphology and systematics of the family Halarachnidae Oudemans 1906 (Acari, Parasitoidea). Bull. Bingham Oceanogr. Coll. 10:235–266.

Nieman, C., and Klein Obbink, H. J. 1954. The biochemistry and pathology of hypervitaminosis A [vol. 12, p. 69–100]. *In* Vitamins and hormones, advances in research and applications. New York, Academic Press.

Nikulin, P. G. 1937. [Sea-lions of the Okhotsk Sea and their exploitation]. Bull. Pacif. Sci. Inst. Fish., Vladivostock 10:35–48. (In Russian, with English summary.)

Nikulin, P. G. 1940. [Description of the Chukotsk walrus, its habits, distribution in the Chukchi and Bering seas, its food and migrations . . .]. Bull. Pacif. Sci. Inst. Fish., Vladivostock 20:21–59. (In Russian; English transl. in British Columbia Univ. Dept. Zool.)

Nilsson, S. 1820. Däggande djuren [vol. 1, 66 + 419 p.]. In Skandinavisk fauna, en handbok for jägare och zoologer. Lund, 3 vols., 1820–42.

Nilsson, S. 1841. Entwurf einer systematischen Eintheilung und speziellen Beschreibung der Phoken . . . [transl. by W. Peters]. Arch. Naturgesch., Jahrg. 7, Band 1:301–333.

Nordquist, O. 1899. Beitrag zur Kenntnis der isolierten Formen der Ringelrobbe (*Phoca foetida* Fabr.). Acta Soc. Fauna Flora Fenn. 15:1–44, 1 tbl., 3 pls.

Norton, A. H. 1930. The mammals of Portland, Maine, and vicinity. Proc. Portland Soc. Nat. Hist. 4: 1–149, fold. map.

Nybelin, O. 1931. Säugetier-und Vogelcestoden von Juan Fernandez [vol. 3, zoology, p. 493–523, 2 pls.]. *In* The natural history of Juan Fernandez and Easter Island, ed. by Dr. Carl Skottsberg. Uppsala, Almqvist and Wiksells, 1921–40.

Ognev, S. I. 1935. [Pinnipedia, vol. 3, p. 316–600]. *In* The mammals of eastern Europe and northern Asia [title varies]. Moscow and Leningrad, State Publishing House, vol. 1 (1928) to vol. 7 (1950). (In Russian.)

Okada, Y. 1938. A catalogue of vertebrates of Japan. Tokyo, Maruzen, 4 + 412 p. (In English and Japanese.)

Olds, J. M. 1950. Notes on the hood seal (*Cystophora cristata*). J. Mammal. 31:450–452.

Oliver, W. R. B. 1921. The crab-eating seal in New Zealand. Trans. Proc. N. Z. Inst. 53:360, 1 pl.

Orlov, J. A. 1931. Über die Reste eines primitiven Pinnipedier aus den neogenen Ablagerungen Westsibiriens. C. R. Acad. Sci. U.R.S.S. ser. A, no. 3:67–70.

Ortmann, A. E. 1897. [Distribution of marine mammals]. Science, n.s., 5:957–958.

Osgood, W. H. 1904. A biological reconnaissance of the base of the Alaska Peninsula. N. Amer. Fauna 24:1–86, 7 pls.

Osgood, W. H. 1943. The mammals of Chile. Publ. Field Mus., zool. ser., 30:1–268.

Pallas, P. S. 1811–42. Phocae [vol. 1, p. 99–119]. *In* Zoographia Rosso-Asiatica, sistens omnium animalium in extenso imperio Rossico et adjacentibus maribus . . . Petropoli, in officina Caes. Academiae Scientarum impress, 3 vols.

Palmer, R. S. 1954. Seals and sea lions. Order Pinnipedia [p. 148–166, 2 col. pls.]. *In* The mammal guide . . . Garden City, N.Y., Doubleday.

Palmer, T. S. 1904. Index generum mammalium. A list of the genera and families of mammals. N. Amer. Fauna 23:1–984.

Paulian, P. 1952. Sur la présence aux Iles Kerguelen d'*Hydrurga leptonyx* (Bl.) et d'*Arctocephalus gazella* (Pet.) et notes biologiques sur deux phocidés. Mammalia 16:223–227.

Paulian, P. 1953. Pinnipèdes, cétacés, oiseaux des Iles Kerguelen et Amsterdam. Mission Kerguelen 1951. Mém. Inst. Sci. Madagascar, sér. A, Tome 8:111–234, 30 pls.

Paulian, P. 1956. Exploitation, destruction et protection des pinnipèdes. La Terre et la Vie, no. 1, p. 1–10 (extrait).

Paulian, P. 1957*a*. Note préliminaire sur la systématique de l'otarie de l'Ile Amsterdam. Mammalia 21:9–14.

Paulian, P. 1957*b*. Note sur les phoques des Iles Amsterdam et Saint-Paul. Mammalia 21:211–225, 2 pls.

Peale, T. R. 1848. Mammalia and ornithology [vol. 8, 17 + 338 p.]. *In* United States exploring expedition during the years 1838 . . . 1842, under the command of Charles Wilkes, U.S.N. Philadelphia, C. Sherman.

Pennant, T. 1771. Synopsis of quadrupeds. Chester, 25 + 382 p., 31 pls.

Pennant, T. 1781. History of quadrupeds. London, 2 vols.

Perkins, J. 1945. Biology at Little America III, the west base of the United States antarctic service expedition 1939–1941. Proc. Amer. Phil. Soc. 89: 270–284.

Pernetty, Dom A. J. 1770. Histoire d'un voyage aux Isles Malouines, fait en 1763 et 1764 . . . Paris, Saillant et Nyon, Delalain, 2 vols. (Spelled Pernety in English version publ. 1741.)

Péron, F. 1807–16. Voyage de découvertes aux terres australes, exécuté . . . pendant les années 1800–1804. Paris, Imprimerie Royale, 2 vols. and atlas. (Vol. 2, at least, is posthumous, 1816, ed. by Louis Freycinet. The collaboration of Charles-Alexandre Leseuer, natural history artist of the expedition, in zoological phases of the report is acknowledged. Allen, 1939, p. 249, gave 1817 as terminal date.)

Peters, W. C. H. 1866*a*. Über die Ohrenrobben (Seelöwen und Seebaren), Otariae, insbesondere über die in den Sammlungen zu Berlin benfindlichen Arten. Monatsber. K. P. Akad. Wissensch. Berlin 1866:261–281, 4 pls.

Peters, W. C. H. 1866*b*. Nachtrag zu seiner Abhandlung über die Ohrenrobben (Otariae). Monatsber. K. P. Akad. Wissensch. Berlin 1866:665–672, 1 pl.

Peters, W. C. H. 1871. Über eine für Chile neue Art von *Otaria* . . . Monatsber. K. P. Akad. Wissensch. Berlin 1871:558–566, 2 pls. (See also Philippi, 1871.)

Peters, W. C. H. 1875. Über eine neue Art von Seebären, *Arctophoca gazella*, von der Kerguelen-Inseln. Monatsber. K. P. Akad. Wissensch. Berlin 1875: 393–399.

Peters, W. C. H. 1876. Über die Pelzrobbe von der Inseln St. Paul und Amsterdam . . . Monatsber. K. P. Akad. Wissensch. Berlin 1876:315–316.

Peters, W. C. H. 1877. Über die Ohrenrobben, Otariae, als Nachtrag zu seiner im vorigen Jahre über diese Thiere gelesenen Abhandlung. Monatsber. K. P. Akad. Wissensch. Berlin 1877:505–507.

Philippi, R. A. 1871. [Communication, p. 558–562, describing *Otaria argentata* n. sp. *In* Peters, 1871].

Pohle, H. 1927. Die Pinnipedier . . . [vol. 19, p. 449–462]. *In* Deutsche Südpolar-Expedition 1901–1903, ed. by Erich von Drygalski. Berlin, W. deGruyter, 20 vols., 1921–31.

Pohle, H. 1932. Die Säugetiere des arktischen Gebietes [Band 6, Leif. 2, p. 67–80]. *In* Fauna arctica . . . Jena, G. Fischer, 6 vols., 1900–1933.

Poole, A. J., and Viola S. Schantz. 1942. Catalog of the type specimens of mammals in the United States National Museum, including the Biological Surveys collection. Bull. U.S. Nat. Mus. 178, 14 + 705 p.

Prosser, C. L. 1950. Water [chap. 2, p. 6–74]. *In* Comparative animal physiology, C. L. Prosser, ed. Philadelphia, W. B. Saunders.

Quoy, J.-R.-C., and J.-P. Gaimard. 1830. Pinnipedia [vol. 1, p. 89–99]. *In* Voyage de la corvette "l'Astrolabe" exécuté . . . pendant les années 1826 . . . 1829, sous le commandement de J. Dumont d'Urville . . . Paris, 15 vols. in 12, 1830–35; atlas 5 vols. 1833–34.

Rand, R. W. 1950*a*. On the milk dentition of the Cape fur seal. Off. J. Dent. Ass. S. Afr. (15 October 1950), 10 p. (Separate seen.)

Rand, R. W. 1950*b*. Studies on the Cape fur-seal (*Arctocephalus pusillus,* Schreber). Union S. Africa Dept. Agric., Govt. Guano Islands Admin., Progr. Repts. 1–3, *circa* 60 processed p. with interleaved printed tbls. and figs.

Rand, R. W. 1956*a*. Notes on the Marion Island fur seal. Proc. Zool. Soc. Lond. 126:65–82.

Rand, R. W. 1956*b*. The Cape fur seal *Arctocephalus pusillus* (Schreber); its general characteristics and moult. Commerce and Industry, Union S. Africa Div. Fish., Investigational Rept. 21, 52 p. (Separate seen.)

Rasmussen, B. 1957. Exploitation and protection of the East Greenland seal herds. Norsk Hvalfangsttid. 1957:45–49. (Norwegian and English.)

Rass, T. S., A. G. Kaganovskiy, and S. K. Klumov, editors. 1955. [Pinnipedia, sect. 4, p. 95–115]. *In* Geographical distribution of fishes and other commercial animals of the Okhotsk and Bering Seas. Trudy Instituta Okeanologii, Tom 14, p. 1–120. (In Russian.)

Rausch, R., B. B. Babero, R. V. Rausch, and E. L. Schiller. 1956. Studies on the helminth fauna of Alaska. 27. The occurrence of larvae of *Trichinella spiralis* in Alaskan mammals. J. Parasit. 42:259–271.

Richter, C. P., and H. D. Mosier, Jr. 1954. Maximum sodium chloride intake and thirst in domesticated and wild Norway rats. Amer. J. Physiol. 176:213–222.

Roberts, A. 1951. The mammals of South Africa. Johannesburg, "The Mammals of South Africa Book Fund," 48 + 700 p.

Roberts, B. B. 1948. Chronological list of antarctic expeditions with brief notes on each, 1502–1948 [p. 6–25]. *In* The antarctic pilot. Ed. 2, London, Admiralty, Hydrographic Dept., 42 + 371 p., maps.

Rothschild, W. 1908. *Mirounga angustirostris* (Gill). Novit. Zool. 15:393–394, 8 pls.

Rothschild, W. 1910. Notes on sea elephants (*Mirounga*). Novit. Zool. 17:445–446, 2 pls.

Rudmose Brown, R. N. 1913. The seals of the Weddell Sea: notes on their habits and distribution. Edinburgh, Scottish Natl. Antarctic Exped., Sci. Res. Voyage "Scotia," 1902–4, vol. 4 (zool.), pt. 13, p. 181–198. 9 pls.

Rüe, E. A. de la. 1950. Notes sur les Iles Crozet. Bull. Mus. Hist. Nat., Paris 22: 197–203.

Saemundsson, B. 1939. Pinnipedia [p. 6–14]. *In* Mammalia. The zoology of Iceland, v. 4, part 76, p. 1–52. Copenhagen and Reykjavik, Ejnar Munksgaard.

Santiago Carrara, I. 1952. Lobos marinos, pingüinos y guaneras de las costas del litoral marítimo e islas adyacentes de la República Argentina. Universidad Nac. de La Plata, Facultad de Ciencias Veterinarias, Catedra de Higiene e Industrias, Publ. Especial, 189 p., 1 map, 9 p. appendix.

Santiago Carrara, I. 1954. Observaciones sobre el estado actual de las poblaciones de pinnipedos de la Argentina. Eva Peron, Argentina [privately published?], 17 p., partly processed.

Sapin-Jaloustre, J. 1952–53. Les phoques de Terre Adélie. Mammalia, 16:179–212, 5 pls.; 17:1–20, 2 pls.

Schauinsland, H. H. 1899. Drei Monate auf einer Koralleninsel (Laysan). Bremen, Max Nössler, 104 p.

Scheffer, V. B. 1942. A list of the marine mammals of the west coast of North America. Murrelet, Seattle, 23:42–47.

Scheffer, V. B. 1945. Growth and behavior of young sea lions. J. Mammal. 26: 390–392.

Scheffer, V. B. 1948. Use of fur-seal carcasses by natives of the Pribilof Islands, Alaska. Pac. Northw. Quart. 39:131–133.

Scheffer, V. B. 1950*a*. Probing the life secrets of the Alaska fur seal. Pacific Discovery 3:22–30.

Scheffer, V. B. 1950*b*. The food of the Alaska fur seal. U.S. Dept. Interior, Fish and Wildlife Service, Wildl. Leafl. 329, 16 p.

Scheffer, V. B. 1950*c*. Growth layers on the teeth of Pinnipedia as an indication of age. Science 112:309–311.

Scheffer, V. B. 1950*d*. Winter injury to young fur seals on the northwest coast. Calif. Fish Game 36:378–379.

Scheffer, V. B. 1951. Cryptorchid fur seals. Amer. Midl. Nat. 46:646–648.

Scheffer, V. B. 1955. Body size with relation to population density in mammals. J. Mammal. 36:493–515.

Scheffer, V. B. 1956. Little-known reference to name of a harbor seal. J. Wash. Acad. Sci. 46:352.

Scheffer, V. B., N. L. Karrick, and F. B. Sanford. 1950. Vitamin A in selected, pale-colored livers of Alaska fur seals, 1948. U.S. Dept. Interior, Fish and Wildlife Service, Spec. Sci. Rept., Fisheries 32, 8 p.

Scheffer, V. B., and J. W. Slipp. 1944. The harbor seal in Washington state. Amer. Midl. Nat. 32:373–416.

Schenk, E. T., and J. H. McMasters. 1956. Procedure in taxonomy . . . Ed. 3, enlarged and in part rewritten by A. Myra Keen and S. W. Muller. Stanford Univ. Press, 7 + 119 p.

Schlegel, H. (see Temminck, 1847).

Scholander, P. F. 1955. Evolution of climatic adaptation in homeotherms. Evolution 9:15–26.

Scholander, P. F., V. Walters, R. Hock, and L. Irving. 1950*a*. Body insulation of some arctic and tropical mammals and birds. Biol. Bull. 99:225–236.

Scholander, P. F., V. Walters, R. Hock, and L. Irving. 1950*b*. Adaptation to cold in arctic and tropical mammals and birds in relation to body temperature, insulation, and basal metabolic rate. Biol. Bull. 99:259–271.

Schreber, J. C. D. von. [1774]–1846. Die Säugethiere in Abbildungen nach der Natur, mit Beschreibungen. Leipzig, 7 parts, col. pls.

Schwartz, E. 1942. The harbor seal of the western Pacific. J. Mammal. 23: 222–223.

Sclater, P. L. 1897–98. On the distribution of marine mammals. Science, n.s., 5:956–957 (1897); Proc. Zool. Soc. Lond. 1897:349–359, 1 map (1898).

Scopoli, G. A. 1777. Introductio ad historiam naturalem, sistens genera lapidum, plantarum et animalium . . . Pragae, 10 + 506 + 34 p.

Scott, H. H., and C. Lord. 1926. Studies in Tasmanian mammals, living and extinct. Nos. 13 and 14. The eared seals of Tasmania. Pap. Roy. Soc. Tasm. 1925:75–78, 187–194, pls. 16–21.

Serventy, D. L. 1948. A record of the leopard seal in western Australia. W. Aust. Nat., Perth, 1:155.

Seton, E. T. 1923. The arctic prairies . . . New York, Scribner's Sons, 12 + 308 p.

Shaw, G. 1800. Seals [vol. 1, part 2, p. 249–273, 4 pls.]. *In* General zoology or systematic natural history. London, G. Kearsley, 14 vols. in 28, 1800–1826.

Sherborn, C. D. 1902–33. Index animalium sive index nominum quae ab A. D. [1758] generibus et speciebus animalium imposita sunt. Cambridge Univ. Press, sect. 1, vol. 1, 1902. Brit. Mus. (Nat. Hist.) sect. 2, 33 parts, 1922–33.

Siebold, P. F. von (see Temminck, 1847).

Simpson, G. G. 1945. The principles of classification and a classification of mammals. Bull. Amer. Mus. Nat. Hist. 85, 16 + 350 p.

Sivertsen, E. 1941. On the biology of the harp seal . . . Hvalråd. Skr. 26, 10 + 166 p., 11 pls.

Sivertsen, E. 1953*a*. A new species of sea lion, *Zalophus wollebaeki*, from the Galapagos Islands. K. Norske Vidensk. Selsk. Forh. 26:1–3.

Sivertsen, E. 1953*b*. A review of the family of Otariidae. (One-page separate headed "XIV. Intern. Zool. Cong. Copenhagen 1953," no page number.)

Sivertsen, E. 1954. A survey of the eared seals (Family Otariidae) with remarks on the antarctic seals collected by M/K "Norvegica" in 1928–1929. Det Norske Videnskaps-Akademi i Oslo, Sci. Results Norweg. Antarct. Exped. 1927–1928 et sqq . . . Lars Christensen, no. 36, 76 p., 10 pls. [4 May 1954].

Skinner, J. S. 1957. Seal finger. Amer. Med. Assoc. Arch. Dermatol. 75:559–561.

Slevin, J. R. 1935. An equatorial wonderland. The Galápagos . . . Nat. Hist., N.Y. 36:374–388.

Slijper, E. J. 1956. Some remarks on gestation and birth in Cetacea and other aquatic mammals. Hvalråd. Skr. 41:1–62.

Smirnov, N. A. 1908. [Review of Russian pinnipeds]. Mém. Acad. Sci. St.-Petersb., ser. 8 (phys.-math.), vol. 23., no. 4, p. 1–75, 1 pl. (In Russian.)

Smirnov, N. A. 1927. Diagnostical remarks about some seals (Phocidae) of the northern hemisphere. Trømso. Mus. Aarsh. 48 (1925), no. 5, p. 1–23.

Smirnov, N. A. 1929. Diagnoses of some geographical varieties of the ringed seal (*Phoca hispida* Schreb.). C. R. Acad. Sci. U.R.S.S. 1929, A, no. 4, p. 94–96.

Smith, J. M., and R. J. G. Savage. 1956. Some locomotory adaptations in mammals. J. Linn. Soc. Lond. 62:603–622.

Soper, J. D. 1928. A faunal investigation of southern Baffin Island. Bull. Nat. Mus. Canada 53:1–143, 7 pls.

Soper, J. D. 1944. The mammals of southern Baffin Island, Northwest Territories, Canada. J. Mammal. 25:221–254.

Sorensen, J. H. 1950. Elephant seals of Campbell Island. N. Z. Dept. Sci. Industr. Res., Wellington, Cape Exped. Series, Bull. 6, 31 p.

Spärck, R. 1956. Protection of arctic animals. [p. 55–56]. *In* Proc. and Pap. 5th Technical Meeting, Internatl. Union for the Protection of Nature, Copenhagen, 1954. Bruxelles, pub. by the Secretariat.

Spindler, M., and E. Bluhm. 1934. Kleine Beiträge zur Psychologie des Seelöwen (*Eumetopias californianus*) [=*Zalophus*]. Z. Vergl. Physiol. 21: 616–631.

Starks, E. C. 1922. Records of the capture of fur seals on land in California. Calif. Fish Game 8:155–160.

Steenstrup, J. J. S. 1860. Om Walrossen. Öfvers. Vetensk. Akad. Forh., Stockh., sextonde årgången, 1859: 441–447.

Stefansson, V. 1943. The friendly arctic. New York, Macmillan, 38 + 812 p., pls. and map.

Stefansson, V. 1945. Arctic manual. New York, Macmillan, 16 + 556 p.

Stejneger, L. 1914. The systematic name of the Pacific walrus. Proc. Biol. Soc. Wash. 27:145.

Stejneger, L. 1936. Georg Wilhelm Steller. The pioneer of Alaskan natural history. Harvard Univ. Press, 623 p., 29 pls.

Steller, G. W. 1751. De Bestiis marinis. Novi Comm. Acad. Sci. Petropolitanae, vol. 2, p. 289–398, 3 pls. (English transl. by Walter Miller and Jennie Emerson Miller [part 3, p. 179–218]. *In* The fur seals and fur-seal islands . . . See Jordan and Clark, 1898–1899.)

Storr, G. C. C. 1780. Prodromus methodi mammalium . . . Tübingen, Frid. Wolffer, 43 p., 4 tbls.

Strays, W. L., Jr. 1956. Retia mirabilia of cetaceans. Science 124:167.

Tabulae Biologicae (many editors). 1947–51. Oculus. Amsterdam and Den Haag, W. Junk, pars 1 (1947), pars 2 (1951).

Taylor, F. H. C., M. Fujinaga, and F. Wilke. 1955. Distribution and food habits of the fur seals of the North Pacific Ocean . . . U. S. Dept. Interior, Fish and Wildlife Service, 10 + 86 p.

Temminck, C. J. [1847]. Les mammifères marins [p. 1–26, pls. 21–23]. *In* Fauna Japonica . . . by P. F. von Siebold *et al.* Lugduni Batavorum, A. Arnz et socios, 6 vols., 1833–50. (For date see Catalogue Libr. Zool. Soc. Lond., ed. 5, 1902, p. 617. The vols. are apparently unnumbered. The vol. on mammals seen by the writer in Cambridge Univ. has two parts: "Aperçu général et spécifique sur les mammifères . . .," p. 1–59, pls. 1–20, and "Les mammifères marins," p. 1–26, pls. 21–23. In the "aperçu," p. 6, Temminck acknowledges the help of "mon collaborateur M. Schlegel" in preparation of the section on marine mammals. Hence, Temminck and Schlegel are often credited as joint authors, though only Temminck's name appears at the head.)

Themido, A. A. 1947. As focas das costas de Portugal. Mem. Mus. Zool. Univ. Coimbra 179:1–12, 1 pl.

Thenius, E. 1949. Über die systematische und phylogenetische Stellung der Genera *Promeles* und *Semantor*. S. B. Öst. Akad. Wiss., Abt. 1, Band 158, Heft 4:323–335.

Thomas, O. 1911. The mammals of the tenth edition of Linnaeus; an attempt to fix the types of the genera and the exact bases and localities of the species. Proc. Zool. Soc. Lond. 1911:120–158.

Thompson, S. H. 1951. Seal fisheries [p. 716–732]. *In* Marine products of commerce . . . ed. by D. K. Tressler and J. McW. Lemon. New York, Reinhold, 14 + 782 p.

Thunberg, C. P. 1811. Mammalia capensia, recensita et illustrata. Mém. Acad. Sci. St.-Pétersb., Tome 3:299–323.

Tilesius, W. G. 1835. Die Wallfische. Oken's Isis, 709–752.

Till, W. M. 1954. Mites endoparasitic in the respiratory tract of the Cape sea lion [fur seal]. J. Ent. Soc. S. Afr. 17:266–267.

Troitzky, A. 1953. Contribution a l'étude des pinnipèdes à propos de deux phoques de la Méditerranée . . . Bull. Inst. Océanogr. Monaco, vol. 50, whole no. 1032, 2 + 46 p.

Troll-Obergfell, B. 1928. Der Haarwechsel des See-Elefanten. Anat. Anz. 65: 327–333.

Troll-Obergfell, B. 1930. Über einege Besonderheiten der Haare des gemeinen Seehundes (*Phoca vitulina* L.). Anat. Anz. 69:404–415.

Trouessart, E.-L. 1881. Du rôle des courants marins dans la distribution géographiques des mammifères amphibies, et particulièrement des otaries. C. R. Acad. Sci., Paris 92:1118–1121.

Trouessart, E.-L. 1897–1905. Catalogus mammalium tam viventium quam fossilium. Berlin, R. Friedländer und Sohn. Tomus 1 (1897), fascic. 1, p. 1–218; fascic. 2, p. 219–452; fascic. 3, p. 453–664. Tomus 2 (1898), fascic. 4, p. 665–998; fascic. 5, p. 999–1264; fascic. 6, p. 1265–1469. Quinquennale supplementum (1904), fascic. 1–2, p. 1–546; (1905), fascic. 3–4, p. 547–929.

Trouessart, E.-L. 1907. Mammifères pinnipèdes [p. 1–28, pls. 1–4]. *In* Expedition antarctique Française (1903–1905) comandée par le Dr. Jean Charcot. Sciences naturelles: documents scientifiques. Paris, Masson, 132 p., 19 pls.

Trouessart, E.-L. 1922. La distribution géographique des animaux. Paris, Octave Doin, 12 + 332 p. (One vol. of Encyclopédie Scientifique, Bibliothèque de Zoologie; semipopular.)

Troughton, E. 1951. Furred animals of Australia. Ed. 4. Sydney, Angus and Robertson, 32 + 376 p, 25 pls.

True, F. W. 1906. Description of a new genus and species of fossil seal from the Miocene of Maryland. Proc. U.S. Nat. Mus. 30:835–840, 2 pls.

Turbott, E. G. 1949. Observations on the occurrence of the Weddell seal in New Zealand. Rec. Auckland [N.Z.] Inst. 3:377–379, 1 pl.

Turbott, E. G. 1952. Seals of the Southern Ocean [p. 195–215]. *In* The Antarctic today. A mid-century survey by the New Zealand Antarctic Society. A. H. Wellington and A. W. Reed in conjunction with the Society, 389 p., fold. map.

Turner, W. 1888. Report on the seals collected during the voyage of H.M.S. "Challenger" in the years 1873–76 [vol. 26, no. 2, p. 1–240, 10 pls.; Challenger zool. repts., part 68]. *In* Report on the scientific results of the voyage of H.M.S. "Challenger". . . Edinburgh, Challenger Office, 1880–95.

Turner, W. 1912. The marine mammals in the anatomical museum of the University of Edinburgh. London, Macmillan, 15 + 207 p., 17 pls.

Turton, W. 1806. A general system of nature through the three grand kingdoms . . . by Sir Charles Linné . . . London. Lackington, Allen and Co. 7 vols.

United States Senate. 1957. Interim convention on conservation of North Pacific fur seals . . . 85th congr., 1st sess., executive J, 18 p.

Vaz Ferreira, R. 1950. Observaciones sobre la Isla de Lobos. Montevideo, Univ. Repub. Uruguay, Rev. Fac. Human. Cienc., 5:145–176.

Vaz Ferreira, R. 1956*a*. Características generales de las islas Uruguayas habitadas por lobos marinos. Montevideo, Servicio Oceanografico y de Pesca, Trabajos Sobre Islas de Lobos y Lobos Marinos, no. 1, 21 p.

Vaz Ferreira, R. 1956*b*. Etología terrestre de *Arctocephalus australis* (Zimmermann) ("lobo fino") en las islas Uruguayas. Montevideo, Servicio Oceanografico y de Pesca, Trabajos Sobre Islas de Lobos y Lobos Marinos, no. 2, 22 p.

Venables, U. M., and L. S. V. Venables. 1955. Observations on a breeding colony of the seal *Phoca vitulina* in Shetland. Proc. Zool. Soc. Lond. 125: 521–532.

Verrill, A. E. 1870. [Review of Allen, 1870]. Amer. J. Sci., ser. 2, no. 50, p. 431.

Vibe, C. 1956. The walrus west of Greenland [p. 79–84]. *In* Proc. and Pap. 5th Technical Meeting, Internatl. Union for the Protection of Nature, Copenhagen, 1954. Bruxelles, pub. by the Secretariat.

Vinson, J. 1956. Sur la présence de l'éléphant de mer aux Mascareignes. Proc. Roy. Soc. Arts Sci. Mauritius 1:313–318.

Viret, J. 1955. Pinnipèdes fossiles [Tome 17, fasc. 1., p. 336–340]. *In* Traité de zoologie . . . publie sous la direction de Pierre-P. Grassé. Paris, Masson, 18 vols. [to 1955].

Voipio, P. 1956. The biological zonation of Finland as reflected in zootaxonomy. Ann. Zool. Soc. "Vanamo," Helsinki, 18:1–36.

Walbaum, J. J. 1792. Petri Artedi sueci genera piscium . . . Ichthyologiae pars III . . . Grypeswaldiae, Impensis Ant. Ferdin. Röse, 6 + 723 p., 3 pls.

Walls, G. L. 1942. The vertebrate eye and its adaptive radiation. Bloomfield Hills, Michigan, Bull. Cranbrook Inst. Sci. 19, 14 + 785 p.

Weber, M. 1927–28. Die Säugetiere. Jena, Gustav Fischer. Vol. 1, anatomischer Teil (1927) with collab. of H. M. deBurlet. Vol. 2, systematischer Teil (1928) with collab. of Othenio Abel.

Weddell, J. 1825. A voyage towards the south pole, performed in the years 1822–24 . . . London, Longman, 6 + 276 p., 5 maps.

Weed, T. 1936. "Sergeant Finnegan," a sea lion. Nature Mag. 27:343–344.

Wheeler, E. P. 1953. Notes on Pinnipedia [in northern Labrador]. J. Mammal. 34:253–255.

Wilke, F. 1954. Seals of northern Hokkaido. J. Mammal. 35:218–224.

Willett, G. 1943. Elephant seal in southeastern Alaska. J. Mammal. 24:500.

Wilson, E. A. 1907. Mammalia (whales and seals) [p. ix–xii, 1–69, 4 pls.]. *In* National antarctic expedition 1901–04. Natural history. Vol. 2, Zoology (Vertebrata: Mollusca: Crustacea). London, Brit. Mus.

Winge, H. 1941. Pinnipedia [vol. 2, p. 211 and 241–249]. *In* The interrelationships of the mammalian genera, transl. from Danish by E. Deichmann . . . and G. M. Allen . . . Copenhagen, C. A. Reitzel, 3 vols., 1941–42.

Winter, G., and W. Nunn. 1953. The composition of the blubber fat of crabeater seal. J. Sci. Food Agric. 4:439–442.

Wood Jones, F. 1925. The eared seals of South Australia. Rec. S. Aust. Mus. 3:9–16.

Wynne-Edwards, V. C. 1954. Field identification of the common and grey seals. Scottish Nat. 66:192.

Yañez, A. P. 1948. Vertebrados marinos chilenos. I. Mamiferos. Rev. Biol. Mar., Valparaiso 1:103–123.

Zeek, Pearl M. 1951. Double trachea in penguins and sea lions. Anat. Rec. 111:327–344.

Zimmermann, E. A. W. von 1778–83. Geographische Geschichte des Menschen, und der allgemein verbreiteten vierfüssigen Thiere . . . Leipzig, 3 vols.

Zuckerman, S. 1953. The breeding seasons of mammals in captivity. Proc. Zool. Soc. Lond. 122:827–950.

Zukowsky, L. 1914. Über einige seltene und kostbare Tiere in Carl Hagenbecks Tierpark. Zool. Beobachter 55:228–234.

INDEX TO NAMES OF PINNIPEDS

Scientific names are indexed; for vernacular names see table 1 on pages 3–5. Each page number in boldface refers to the main account of a name used in the present scheme of classification. An asterisk calls attention to a new name, or a name used in a new combination.